Audel™

Questions and Answers for Plumbers' Examinations

Audel™
Questions and Answers for Plumbers' Examinations

All New 4th Edition

Rex Miller
Mark Richard Miller
Jules Oravetz

Wiley Publishing, Inc.

Vice President and Executive Group Publisher: Richard Swadley
Vice President and Executive Publisher: Robert Ipsen
Vice President and Publisher: Joseph B. Wikert
Executive Editor: Carol A. Long
Editorial Manager: Kathryn A. Malm
Development Editor: Kevin Shafer
Production Editor: Vincent Kunkemueller
Text Design & Composition: TechBooks

Library of Congress Cataloging-in-Publication Data:

ISBN: 0-7645-6998-8

10   9   8   7   6   5   4   3   2   1

Contents

Foreword

Plumbing and pipefitting play a major role in the construction of every residential, commercial, and industrial building. Of all the building trades, none is as essential to the health and well-being of the community in general, and to the building occupants in particular, as the plumbing trade. It is the obligation and responsibility of every plumber to uphold this vital trust in the installation of plumbing materials and equipment.

All plumbing installations are governed by rules and regulations set forth in plumbing codes that have been adopted from standards established at either the local, state, or national level. In addition, each installation is subject to inspection by a licensed inspector to ensure that all rules and regulations have been completely satisfied.

This book is offered as a guide in the preparation for plumbers' license examinations—for apprentice, journeyman, or master. These examinations are given periodically by local licensing authorities, that is, by municipal, county, state, or other agencies having legal jurisdiction over the licensing and inspection work done by plumbers.

In most localities, a plumber is required by law to secure a license from the enforcing authority in his or her area. This book supplies the license applicant with the required theoretical knowledge and a thorough understanding of the definitions, specifications, and regulations of the recommended minimum requirements for plumbing by the National Bureau of Standards and by a variety of state plumbing codes.

Numerous examples in the form of questions and answers are presented throughout this book, thereby enabling the license applicant to gain a complete knowledge of the types of questions asked in the plumbers' examinations. The importance of careful study to master the fundamental principles underlying each question and answer should be thoroughly impressed on all candidates for licenses in the various grades. Only through this process can the license applicant prepare himself or herself to solve any new or similar problem on the examination.

The interpretations of the various national, state, and local plumbing codes are those of the authors and are not necessarily the official observations of the various plumbing code committees, officials, and representatives.

Rex Miller, Ed.D.
Mark R. Miller, Ph.D.
Jules Oravetz, P.E.

Acknowledgments

No book can be written without the aid of others. It takes a great number of persons to put together, between two covers, the information available about any particular technical trade. The trade of plumbing is no exception. Many firms have contributed to the illustrations and the text of the book. Those involved with the editing and production are listed on the copyright page.

The authors would like to thank everyone involved for his or her contributions. Some of the firms that supplied technical information and illustrations are listed here:

American Standard Co.

Chase Brass and Copper Company

Crane Company

Dunham-Bush, Inc.

Electric Eel Manufacturing Co., Inc.

Eljer Plumbingware, Inc.

General Engineering Company

Genova, Inc.

Gore, W.L. and Associates

Josam Manufacturing Company

Jet, Inc.

Plumb Shop

Ridge Tool Company (Rigid)

Sloan Valve Company, M. Susan Kennedy

About the Authors

Rex Miller was a Professor of Industrial Technology at The State University of New York, College at Buffalo for more than 35 years. He has taught on the technical school, high school, and college level for more than 40 years. He is the author or coauthor of more than 100 textbooks ranging from electronics through carpentry and sheet metal work. He has contributed more than 50 magazine articles over the years to technical publications. He is also the author of seven Civil War regimental histories.

Mark Richard Miller finished his B.S. degree in New York and moved on to Ball State University, where he obtained his master's and went to work in San Antonio. He taught in high school and went to graduate school in College Station, Texas, finishing the doctorate. He took a position at Texas A&M University in Kingsville, Texas, where he now teaches in the Industrial Technology Department as a Professor and Department Chairman. He has coauthored seven books and contributed many articles to technical magazines. His hobbies include refinishing a 1970 Plymouth Super Bird and a 1971 Roadrunner.

Jules A. Oravetz was a professional engineer and the author of numerous books for the plumbing and pipefitting trades, as well as those for building, grounds, and garden maintenance.

Introduction

In a question-and-answer format, the first question would be what is *plumbing?* It is something that has been with us since birth. We have been using it without giving it a thought, that is, unless something goes wrong. This complicated system, like any system, requires periodic time and effort expended on it to maintain and repair system-wide problems. In most cases, these problems are in need of immediate attention. This is where the plumber becomes the key to the proper operation of the more widespread system.

Plumbing can be defined as a system of pipes and fixtures installed in a building for the distribution and use of potable (drinkable) water and the removal of water-borne wastes. It is usually distinguished from water and sewage systems that serve a group of buildings or a city. Each building has its own system of pipes and fixtures designed to serve the purpose assigned to that building.

Historically

The greatest problem of every civilization in which the population has been centralized in cities and towns has been the development of adequate plumbing systems. In certain parts of Europe, the complex aqueducts built by the Romans to supply their cities with potable water can still be seen. However, the early systems built for the disposal of human wastes were less elaborate. Human wastes were often transported from the cities in carts or buckets, or else discharged into an open, water-filled system of ditches and that led from the city to a lake or stream. In rural areas, people dug a hole and used it for the disposal of human waste. Some of these still exist. For privacy, they usually put some kind of shelter, or outhouse, over the hole. However, these outhouses were sometimes built on a hill above the living quarters. Drainage from the waste disposal worked its way to the water supply (well water) that served the family down the hill. This created many cases of diphtheria and other diseases that sickened and killed those using the well water.

Improvements

Plumbing system improvements were very slow in coming. Virtually no progress was made from the time of the Romans until the nineteenth century. Relatively primitive sanitation facilities were inadequate for the large, crowded population centers that sprang up during the Industrial Revolution. This gave way to outbreaks of typhoid fever and dysentery that often spread. The consumption of water contaminated with human wastes was widespread. Eventually, the epidemics were curbed by the development of separate, underground water and sewage systems that eliminated open sewage ditches. In addition, plumbing fixtures were designed to handle potable water and water-borne wastes within buildings.

Modern Systems

Methods of water distribution vary. For towns and cities, municipally or privately owned water companies treat and purify water collected from wells, lakes, rivers, and ponds, and distribute it to individual buildings. In rural areas, water is commonly obtained directly from individual wells.

In most cities, water is forced through the distribution system by pumps. In rare instances, the source of water is located in mountains or hills above a city, and the pressure generated by gravity is sufficient to distribute water throughout the system. In other cases, water is pumped from the collection and purification facilities into elevated storage tanks and then allowed to flow throughout the system by gravity. But in most municipalities, water is pumped directly through the system. Elevated storage tanks may also be provided to serve as pressure-stabilization devices and as an auxiliary source in the event of pump failure or of a catastrophe (such as fire) that might require more water than the pumps or the water source are able to supply.

Water Pressure

The pressure developed in the water supply system and the friction generated by the water moving through the pipes are the two factors that limit both the height to which water can be distributed and the maximum flow rate available at any point in the system.

Waste Disposal

A building's system for waste disposal has two parts: the *drainage system* and the *venting system*. The drainage portion comprises pipes leading from various fixture drains to the central main, which is connected to the municipal or private sewage system. The venting system consists of pipes leading from an air inlet (usually the building's roof) to various points within the drainage system. It protects sanitary traps from siphoning or blowing by equalizing the pressure inside and outside the drainage system.

Fixtures

Sanitary fixture traps provide a water seal between the sewer pipes and the rooms in which plumbing fixtures are installed. The most commonly used sanitary trap is a U bend, or dip, installed in the drainpipe adjacent to the outlet of each fixture. A portion of the waste water discharged by the fixture is retained in the U, forming a seal that separates the fixture from the open drainpipes.

A Career as a Plumber

When considering a career as a plumber, keep in mind the following:

- Job opportunities should be excellent, because not enough people are seeking training.
- Most workers learn the trade through 4 or 5 years of formal apprenticeship training.
- Pipelayers, plumbers, pipefitters, and steamfitters make up one of the largest and highest paid construction occupations.

The Plumber's Work

Most people are familiar with plumbers who come to their home to unclog a drain or install an appliance. In addition to these activities, however, plumbers install, maintain, and repair many different types of pipe systems. For example, some systems move water to a municipal water treatment plant and then to residential, commercial, and public buildings. Other systems dispose of waste, provide gas to stoves and furnaces, or supply air-conditioning. Pipe systems in power plants carry the steam that powers huge turbines. Pipes also are used in manufacturing plants to move material through the production process.

Plumbers are considered as being in a trade, generally specializing in one of four areas. *Pipelayers* lay clay, concrete, plastic, or cast-iron pipe for drains, sewers, water mains, and oil or gas lines. Before laying the pipe, pipelayers prepare and grade the trenches either manually or with machines. *Plumbers* install and repair the water, waste disposal, drainage, and gas systems in homes and commercial and industrial buildings. Plumbers also install plumbing fixtures, bathtubs, showers, sinks, and toilets in heating and cooling buildings. They also install automatic controls that are increasingly being used to regulate these systems. Some *pipefitters* specialize in only one type of system. *Steamfitters*, for example, install pipe systems that move liquids or gases under high pressure. *Sprinklerfitters* install automatic fire sprinkler systems in buildings.

Pipelayers, plumbers, pipefitters, and steamfitters use many different materials and construction techniques, depending on the type of project. For example, residential water systems incorporate copper, steel, and plastic pipe that can be handled and installed by one or two workers. Municipal sewerage systems, on the other hand, are made of large cast-iron pipes; installation normally requires crews of pipefitters. Despite these differences, all pipelayers, plumbers, pipefitters, and steamfitters must be able to follow building plans or blueprints and instructions from supervisors, lay out the job, and work efficiently with the materials and tools of the trade. Computers often are used to create blueprints and plan layouts.

When construction plumbers install piping in a house, for example, they work from blueprints or drawings that show the planned location of pipes, plumbing fixtures, and appliances. They first lay out the job to fit the piping into the structure of the house with the least waste of material and within the confines of the structure. They then measure and mark areas in which pipes will be installed and connected. Construction plumbers also check for obstructions such as electrical wiring and, if necessary, plan the pipe installation around the problem.

Sometimes plumbers must cut holes in the walls, ceilings, and/or floors of a house. For some systems, they may hang steel supports from ceiling joists to hold the pipe in place. To assemble a system, plumbers (using saws, pipe cutters, and pipe-bending machines) cut and bend lengths of pipe. They connect lengths of pipe with fittings, using methods that depend on the type of pipe used. For plastic pipe, plumbers connect the sections and fittings with adhesives. For copper pipe, they slide a fitting over the end of the pipe and solder it in place with a torch.

After the piping is in place, plumbers install the fixtures and appliances and connect the system to the outside water or sewer lines. Finally, using pressure gages, they check the system to ensure that the plumbing works properly.

Working Conditions for Plumbers

Because plumbers frequently must lift heavy pipes, stand for long periods, and sometimes work in uncomfortable or cramped positions, they need physical strength as well as stamina. They also may have to work outdoors in inclement weather. In addition, they are subject to possible falls from ladders, cuts from sharp tools, and burns from hot pipes or soldering equipment.

Plumbers engaged in construction generally work a standard 40-hour week. Those involved in maintaining pipe systems (including those who provide maintenance services under contract) may have to work evening or weekend shifts, as well as be on-call. These maintenance workers may spend quite a bit of time traveling to and from work sites.

Details of Employment

Plumbers constitute one of the largest construction occupations, holding about 568,000 jobs in the year 2000. About 52 percent worked for plumbing, heating, and air conditioning contractors engaged in new construction, repair, modernization, or maintenance work. Others did maintenance work for a variety of industrial, commercial, and government employers. For example, those working as pipefitters were employed as maintenance personnel in the petroleum and chemical industries, where manufacturing operations require the moving of liquids and gases through pipes. About one of every seven pipelayers, plumbers, pipefitters, and steamfitters was self-employed.

Jobs for this trade consisting of plumbers, pipelayers, pipefitters, and steamfitters are distributed across the country in about the same proportion as the general population.

Training, Other Qualifications, and Advancement

Virtually all pipelayers, plumbers, pipefitters, and steamfitters undergo some type of apprenticeship training. Many programs are administered by local union-management committees made up of members of the United Association of Journeymen and Apprentices of the Plumbing and Pipefitting Industry of the United States and Canada, and local employers who are members of the Mechanical Contractors Association of America, the National Association of Plumbing-Heating-Cooling Contractors, or the National Fire Sprinkler Association.

Nonunion training and apprenticeship programs are administered by local chapters of the Associated Builders and Contractors, the National Association of Plumbing-Heating-Cooling Contractors, the American Fire Sprinkler Association, or the Home Builders Institute of the National Association of Home Builders.

Apprenticeships (both union and nonunion) consist of 4 or 5 years of on-the-job training, in addition to at least 144 hours per year of related classroom instruction. Classroom subjects include drafting and blueprint reading, mathematics, applied physics, chemistry, safety, and local plumbing codes and regulations. On the job, apprentices first learn basic skills, such as identifying grades and types of pipe, using the tools of the trade, and safely unloading materials. As apprentices gain experience, they learn how to work with various types of pipe and how to install different piping systems and plumbing fixtures. Apprenticeship gives trainees a thorough knowledge of all aspects of the trade. Although most pipelayers, plumbers, pipefitters, and steamfitters are trained through apprenticeship, some still learn their skills informally on the job.

Applicants for union or nonunion apprentice jobs must be at least 18 years old and in good physical condition. Apprenticeship committees may require applicants to have a high school diploma or its equivalent. Armed Forces training in pipelaying, plumbing, and pipelining is considered very good preparation. In fact, persons with this background may be given credit for previous experience when entering a civilian apprenticeship program. Secondary or postsecondary courses in shop, plumbing, general mathematics, drafting, blueprint reading, computers, and also physics are good preparation.

Although there are no uniform national licensing requirements, most communities require plumbers to be licensed. Licensing requirements vary from area to area, but most localities require workers to pass an examination that tests their knowledge of the trade and of local plumbing codes.

Some pipelayers, plumbers, pipefitters, and steamfitters may become supervisors for mechanical and plumbing contractors; others go into business for themselves.

Job Outlook for Plumbers

Job opportunities are expected to be excellent, as increased demand for skilled pipelayers, plumbers, pipefitters, and steamfitters is expected to outpace the supply of workers trained in this craft. Employment of pipelayers, plumbers, pipefitters, and steamfitters is expected to grow about as fast as the average for all occupations through the year 2010. In addition, many potential workers may prefer work that is less strenuous and has more comfortable working conditions. Well-trained workers will have especially favorable opportunities.

Demand for plumbers will stem from building renovation, including the increasing installation of sprinkler systems; repair and maintenance of existing residential systems; and maintenance activities for places having extensive systems of pipes (such as power plants, water and wastewater treatment plants, pipelines, office buildings, and factories). However, the growing use of plastic pipe and fittings (which are much easier to install and repair than other types) increasingly efficient sprinkler systems, and other new technologies will restrict the number of new jobs. In addition to those resulting from employment growth, many positions will become available each year from the need to replace experienced workers who retire, die, or leave the occupation for other reasons.

Traditionally, many organizations with extensive pipe systems have employed their own plumbers to maintain equipment and keep systems running smoothly. But, to reduce labor costs, many of these firms no longer employ a full-time, in-house plumber. Instead, when they need a plumber, they rely on workers provided under service contracts by plumbing contractors.

Construction projects provide only temporary employment. So, when a project ends, plumbers working on the project may experience bouts of unemployment. Because construction activity varies from area to area, job openings, as well as apprenticeship opportunities, fluctuate with local economic conditions. However, employment of pipelayers, plumbers, pipefitters, and steamfitters generally is less sensitive to changes in economic conditions than is that of some other construction trades. Even when construction activity declines, maintenance, rehabilitation, and replacement of existing piping systems, as well as the increasing installation of fire sprinkler systems, provide many jobs for plumbers.

Plumber Earnings

Plumbers, pipelayers, pipefitters, and steamfitters are among the highest paid construction occupations; in the year 2000, median hourly earning of plumbers, pipefitters, and steamfitters was $18.19. The middle 50 percent earned between $14.00 and $24.24. The lowest 10 percent earned less than $10.71, and the highest 10 percent earned more than $30.06. Median hourly earnings in the industries employing the largest numbers of plumbers in 2000 are shown here:

Plumbing, heating, and air-conditioning	$18.20
Nonresidential building construction	$17.80
Heavy construction, except highway	$17.26
Local government	$17.12
Miscellaneous special trade contractors	$16.92

In 2000, median hourly earning of plumbers was $13.20. The middle 50 percent earned between $10.17 and $17.71. The lower 10 percent earned less than $8.61, and the highest 10 percent earned more than $23.16.

Apprentices usually begin at about 50 percent of the wage rate paid to experienced plumbers. Wages increase periodically as skills improve. After an initial waiting period, apprentices receive the same benefits as do experienced plumbers.

Many plumbers are members of the United Association of Journeymen and Apprentices of the Plumbing and Pipefitting Industry of the United States and Canada.

Occupations Related to Plumbing

Other occupations in which workers install and repair mechanical systems in buildings are boilermakers; electricians; elevator installers and repairers; heating, air-conditioning, and refrigeration mechanics and installers; industrial machinery installation, repair, and maintenance workers; sheet-metal workers; and stationary engineers and boiler operators.

Sources of Additional Information on Plumbing

For information about apprenticeships or work opportunities in pipelaying, plumbing, pipefitting, and steamfitting, contact local plumbing, heating, and air-conditioning contractors; a local chapter of the National Association of Plumbing, Heating, and Cooling Contractors; a local chapter of the Mechanical Contractors Association; a local chapter of the United Association of Journeymen and Apprentices in the United States and Canada; or the nearest office of your state employment service or apprenticeship agency.

For information about apprenticeship opportunities for pipelayers, plumbers, pipefitters, and steamfitters, contact the following:

- United Association of Journeymen and Apprentices of the Plumbing and Pipefitting Industry, 901 Massachusetts Ave. NW, Washington, DC 20001.

For more information about training programs for plumbers, pipefitters, and steamfitters, contact the following:

- Associated Builders and Contractors, 1300 N. 17th St., Arlington, VA 22209. (www.abc.org)
- National Association of Home Builders, 15th and M St., NW, Washington, DC 20005. (www.hbi.org)

For general information about the work of pipelayers, plumbers, and pipefitters, contact the following:

- Mechanical Contractors Association of America, 1385 Piccard Dr., Rockville, MD 20850. (www.mcaa.org)
- National Association of Plumbing-Heating-Cooling Contractors, 180 S. Washington St., P.O. Box 6808, Falls Church, VA 22040.

For general information about the work of sprinklerfitters, contact the following:

- American Fire Sprinkler Association, Inc., 12959 Jupiter Rd., Suite 142, Dallas, TX 75238-3200. (www.firesprinkler.org)

- National Fire Sprinkler Association, Robin Hill Corporate Park, Rt. 22, Box 1000, Patterson, NY 12563. (www.nfsa.org)

Studying and Test-Taking

Following is a list of key points to keep in mind when preparing yourself to take a certification exam:

- *Make yourself a study schedule*—Assign yourself a period of time each day to devote to preparing for the exam. A regular time is best. But, the important thing is to study every day.

- *Study alone*—You can concentrate better when you work alone. Keep a list of questions you find puzzling. Mark the points you are unsure of so you can talk them over with a friend who is preparing for the same exam. Then, exchange ideas at a joint review session just before the test.

- *Remove all distractions*—Find a quiet, well-lit spot as far as possible from cell phones, telephones, televisions, and family activities. Arrange not to be interrupted.

- *Start at the beginning*—Read and underline points that you consider important. Make notes on the study page. Mark the pages that you think are important.

- *Pay particular attention to the instructional chapters*—Study the *Plumbing Code Definitions,* the *Dictionary of Plumbing Terms,* and the *Scrambled Dictionary of Equipment and Usage.* Study and learn the language of the field. Pay particular attention to the technique of eliminating wrong answers. This information is important to answering all multiple-choice questions.

- *Be sure to answer all the practice questions chapter by chapter*—Make notes on your weaknesses. Then, use textbooks to brush up.

- *Take previous exams*—When you think that you are ready, move on to the exams that were used in previous tests. If possible, be sure to answer an entire exam in one sitting. However, if you must divide your time, divide it into no more than two sessions per exam.

When taking the practice exams, treat them respectfully. Treat each as a dress rehearsal for the real thing. Time yourself accurately. Do not peek at the correct answer. Remember, you are taking these for practice. They will not be scored. They do not count. So, learn from them.

Important

Do not memorize questions and answers. Any question that has been released to the public will not, in most instances, be used again. Some of the questions may be very similar, but you will not be tested with these *exact* questions. These types of questions will give you good practice, but they will not have the same answers as any of the questions on your exam.

Taking an Exam

- *Arrive at the examination room early*—Getting a good start begins with being familiar with the room. If the room is too cold, too warm, or not well-ventilated, call these conditions to the attention of the person in charge.

- *Read the instructions carefully*—Test-takers lose points because they misread some important part of the directions. (Reading the incorrect choice instead of the correct choice is a good example.)

- *Don't be afraid to guess*—The best policy is to pace yourself, so that you can read and at least consider each question. Sometimes this does not work. However, most civil service exam scores are based only on the number of questions correctly answered. This means that a wild guess is better than a blank space. There is usually no penalty for a wrong answer. You just might guess right. If time is about to run out, mark all the remaining spaces with the same answer. According to the law of averages, some will be right. You have bought this book for practice questions. Part of your preparing for the test is learning to pace yourself so that you need not answer randomly at the end. Of course, far better than a wild guess is an educated guess. You make this kind of guess not when you are pressed for time but when you are not sure of the correct answer. Usually, one or two of the choices are obviously wrong. Eliminate the obviously wrong answers and try to reason among those remaining. Then, if necessary, guess from the smaller field. The odds of choosing a right answer increase if you guess from a field of two instead of from a field of four. When you make an educated guess or a wild guess in the course of the exam, you might want to make a note next to the question number in the test booklet. Then, if there is time, you can go back for a second look.

- *Reason your way through multiple-choice questions*—However, do it carefully and methodically.

Tips on Multiple-Choice Questions
Consider the following sample question:

1. On the job. Your supervisor gives you a hurried set of directions. As you start your assigned task, you realize that you are not quite clear on the directions given to you. The best action to take would be to _____.

   a. continue with your work, hoping to remember the directions and do the best you can.

   b. ask a coworker in a similar position what he or she would do.

   c. ask your supervisor to repeat or clarify certain directions.

   d. go on to another assignment.

In this question you are given four possible answers to the problem described. Though the four choices are all possible actions, it is up to you to choose the *best* course of action in this particular situation.

Choice (a) will likely lead to a poor result. Given that you do not recall or understand the directions, you would not be able to perform the assigned task properly. Keep choice (a) in the back of your mind until you have examined the other alternatives, though, because it could be the best of the four choices given.

Choice (b) is also a possible course of action, but is it the best? Consider that the coworker you consult has not heard the directions. How could he or she know? Perhaps his or her degree of incompetence is greater than yours in this area. Of choices (a) and (b), the better of the two is still choice (a).

Choice (c) is an acceptable course of action. Your supervisor will welcome your questions and will not lose respect for you. At this point, you should hold choice (c) as the best answer and eliminate choice (a).

The course of action in choice (d) is decidedly incorrect, because the job at hand would not be completed. Going on to something else does not clear up the problem; it simply postpones your having to make a necessary decision.

Now that you have made a careful consideration of all the choices given, choice (c) stands out as the best possible course of action. You should select choice (c) as your answer.

Every question is written about a fact or an accepted concept. This sample question indicates the concept that, in general, most supervisory personnel appreciate subordinates questioning directions that may not have been fully understood. This type of clarification precludes subsequent errors on the part of the subordinates. On the other hand, many subordinates are reluctant to ask questions for fear that their lack of understanding will detract from their supervisor's evaluation of their abilities.

The supervisor has the responsibility of issuing orders and directions so that subordinates will be encouraged to ask questions. This is the idea on which the sample question was based.

Of course, if you were familiar with this concept, you would have no trouble answering this question. But, if you were not familiar with it, the method used here for eliminating incorrect choices and selecting the correct one should prove successful for you.

Now that you have seen how important it is to identify the concept and the key phrase of the question, you must understand that equally (or perhaps even more) important is identifying and analyzing the key word or words (the qualifying word) in a question. This word is usually an adjective or adverb. Some of the most common key words are as follows:

always	least	never
average	lowest	only
best	mainly	or
but	maximum	smallest
chiefly	minimum	sometimes
easiest	most	tallest
greatest	most likely	
highest	most nearly	

Knowing these key words is usually half the battle in understanding and answering all types of exam questions.

Next, you could use the elimination method on some additional questions. Consider this example:

2. On the first day you report for work after being appointed as a plumber's helper, you are assigned to routine duties that seem to you to be simple and easy that anyone can do. You should _____.

   a. do your assignment in a routine manner while conserving your energies for more important work in the future.

   b. say to your superior that you are capable of greater responsibility.

   c. see these duties as an opportunity to become thoroughly familiar with the workplace.

    d. try to get someone to take care of your assignment until you have become thoroughly acquainted with your new associates.

Choice (a) will not lead to getting your assigned work done in the best possible manner. It would be your responsibility as a newly appointed plumber's helper to do a good job. The likelihood of getting to do more important work in the future following the approach stated in this choice is remote. However, since this is only choice (a), keep it in mind, because it may turn out to be the best of the four choices given.

Choice (b) is better than choice (a), because your superior may not be familiar with your capabilities at this point. Now you can drop choice (a) and retain choice (b), because, once again, it may be the best of the four choices.

The question clearly states that you are newly appointed. Would it not be wise to perform whatever duties you are assigned in the best possible manner? In this way, you would not only use the opportunity to become acquainted with procedures, but also to demonstrate your abilities. Choice (c) suggests a course of action that will benefit you and the location in which you are working. That is because it will get needed work done. Now drop choice (b) and retain choice (c). It is by far the better of the two choices.

The course of action in choice (d) is not likely to get the assignment completed. And, it will not improve your image with fellow apprentice plumbers. Choice (c), when compared to choice (d), is far better and, therefore, should be selected as the best choice.

Consider another multiple choice question:

3. An off-duty police officer in civilian clothes is riding in the rear of a city bus and notices two teenage boys tampering with the rear emergency door. The most appropriate action for the officer to take would be to _____.

    a. tell the boys to discontinue their tampering, pointing out the dangers to life that their actions may create.

    b. report the boys' actions to the bus operator and let the bus operator take whatever action is deemed best.

    c. signal the bus operator to stop, show the boys his or her officer's badge, and then order them off the bus.

    d. show the boys his or her officer's badge, order them to stop their actions, and take down their names and addresses.

Before considering answers to this question, you must accept the well-known fact that a police officer is always on duty to uphold the law, even though he or she may be technically off-duty.

In choice (a), the course of action taken by the police officer will probably serve to educate the boys and get them to stop their unlawful activity. Since this is only the first choice, we will hold it aside.

In choice (b), you have to realize that the authority of the bus operator in this instance is limited. He can ask the boys to stop tampering with the door, but that is all. The police officer can go beyond that point. Therefore, we drop choice (b) and continue to hold choice (a).

Choice (c) as a course of action will not have a lasting effect. What is to stop the boys from boarding the next bus and continuing their unlawful action? We therefore drop choice (c) and continue to hold choice (a).

Choice (d) may have some beneficial effect, but it would not keep the boys from continuing their actions in the future. When choice (a) is compared with choice (d), it appears that choice (a) is the better one overall, and therefore it is the correct answer.

The next question illustrates a type of question that has gained popularity in recent examinations and that requires a two-step evaluation. First, the reader must evaluate the condition in the question as being desirable or undesirable. Once the determination has been made, you are then left with making a selection from two choices instead of the usual four.

4. A visitor to an office in a city agency tells one of the office clerks that he has an appointment with the supervisor of the office, who is expected back shortly. The visitor asks for permission to wait in the supervisor's private office, which is unoccupied at the moment. For the office clerk to allow the visitor to do so would be _____ .

   a. desirable. The visitor would be less likely to disturb the other employees or *to be* disturbed by them.

   b. undesirable. It is not courteous to permit a visitor to be left alone in an office.

   c. desirable. The supervisor may wish to speak to the visitor in private.

   d. undesirable. The supervisor may have left confidential papers on the desk.

First of all, evaluate the course of action on the part of the office clerk of permitting the visitor to wait in the supervisor's office as being very undesirable. There is nothing mentioned of the nature of the visit. It may be for a purpose that is not friendly or congenial. There may be papers on the supervisor's desk that he or she does not want the visitor to see, or, for that matter, to even know about. Therefore, at this point, you have to decide between choices (b) and (d).

Courtesy is definitely not a question here. All visitors should be treated with courtesy as a general policy, but permitting the visitor to wait in the supervisor's office is not the only possible act of courtesy. Another comfortable place could be found for the visitor to wait.

Choice (d) contains the exact reason for evaluating this course of action as being undesirable, and when compared with choice (b), choice (d) is a far better answer.

Marking Your Answers

You should read all the choices before you mark your answer. It is statistically true that most errors are made when the last choice is the correct answer. Too many people mark the first answer that seems correct without reading through all the choices to find out which answer is *best*.

Read the following suggestions and review them before you take the actual exam. Become familiar with the suggestions. This will make you feel more comfortable with the exam and you will find them all useful when you are marking your choices.

- Mark your answers by completely blackening the answer space of your choice.

- Mark only *one* answer for each question, even if you think that more than one answer is correct. You must choose only one. If you mark more than one answer, the scoring machine will consider you wrong, even if one of your answers is correct.

- If you change your mind, erase completely. Leave no doubt as to which answer you have chosen.

- If you do any figuring on the test booklet or on scratch paper, be sure to mark your answer on the answer sheet.

- Check often to be sure that the question number matches the answer space number and that you have not skipped a space by mistake. If you do skip a space, you must erase all the answers after the skip and answer all the questions again in the right places.

- Answer every question in order, but do not spend too much time on any one question. If a question seems to be impossible, do not take it as a personal challenge. Guess and move on. Remember that your task is to answer correctly as many questions as possible. You must apportion your time so as to give yourself a fair chance to read and answer all the questions. If you guess at an answer, mark the question in the test booklet so that you can find it easily, if time allows.

- Guess intelligently, if you can. If you do not know the answer to a question, eliminate the answers that you know are wrong and guess from among the remaining choices. If you have no idea whatsoever of the answer to a question, guess anyway. Choose an answer other than the first. The first choice is generally the correct answer less often than the other choices. If your answer is a guess (either an educated guess or a wild one), mark the question in the question booklet so that you can give it a second try if time permits.

- If you happen to finish before time is up, check to be sure that each question is answered in the right space and that there is only one answer for each question. Return to the difficult questions that you marked in the booklet and try them again. There is no bonus for finishing early, so use all your time to perfect your exam paper.

With the combination of techniques for studying and test-taking, as well as the self-instructional course and sample examinations in this book, you are given the tools you need to score a high mark on your exam.

Chapter 1

Basic Plumbing Installation Principles

The majority of state plumbing codes and licensing examinations are designed to establish environmental sanitation and safety through properly designed supervision that will ensure properly installed and maintained plumbing systems. Details of plumbing construction vary, but the basic sanitary and safety principles are the same. The desired and required results (to protect the health of people) are similar regardless of locality. These basic principles require that all plumbing in public and private buildings intended for human occupation or use be installed so as to protect the health, welfare, and safety of the occupants and the public.

These basic principles include the following:

- Buildings intended for human occupancy or use will be provided with a supply of pure and wholesome water with connections not subjected to the hazards of backflow or back siphonage and not connected to unsafe water supplies. If there is a public water main available, an individual connection to the public water main shall be made.

- Buildings with plumbing fixtures and devices shall be provided with a supply of water in sufficient volume and pressure to enable them to operate in a satisfactory manner at all times.

- Water heaters and other devices used for purposes of water heating and storing shall be designed and installed to prevent explosion through overheating.

- Where public sanitary sewers are available, buildings intended for human occupancy shall have a connection made to the public sanitary sewer.

- Plumbing fixtures shall be made of materials that are durable, corrosion-resistant, nonabsorbent, and free of concealed fouling surfaces. Rooms in which water closets, urinals, and similar fixtures are installed shall have proper ventilation and adequate lighting.

- It is recommended that family dwelling units adjacent to sanitary sewer lines or having private sewage disposal systems have at least one lavatory, one water closet, a bathtub or shower, and a kitchen-type sink for purposes of sanitation and personal hygiene. Other structures for human occupancy with sanitary public or private sewage-disposal systems should have no less than one water closet and one fixture for hand-washing.

- Building sanitary drainage systems shall be designed, installed, and maintained in a condition so as to conduct wastewater and sewage to designated locations from each fixture with a flow that prevents fouling, clogging, and deposits of solids in the piping. Sufficient cleanout shall be installed so that the piping system can be easily cleaned in case of stoppage.

- Plumbing systems shall be maintained in a sanitary condition, and each connection (direct or indirect) to the drainage system shall have a water-seal trap. The system shall be kept in a serviceable condition with adequate spacing of the fixtures. These fixtures should be reasonably accessible for cleaning.

- Drainage pipe shall be designed and installed with a durable material free of water leakage and offensive odors caused by drain sewer air. Installation shall be in accordance with good workmanship practices and use of good grade material.

- Plumbing systems shall be designed, installed, and kept in adjustment so as to provide the required quantity of water consistent with adequate performance. There should be no undue noise under normal conditions and use. New systems and/or remodeled systems shall be subjected to tests that will disclose leaks and defects.

- Included in the design shall be every consideration for the preservation of the strength of structural members of the building. Each vent terminal extending to the outer air shall be designed to minimize clogging and return of foul air to the building.

- Design considerations shall include protection from contamination by sewage backflow of water, food, disposal of sterilized items, and similar materials. Substances that will clog pipes or their joints and interfere with the sewage disposal process or produce explosive mixtures shall not be allowed in the building sewage drainage system.

- Sewage or other wastes from a plumbing system shall not discharge into subsurface soil or into a water surface unless it has first been treated in an acceptable manner.

The following are typical questions asked in reference to basic plumbing and installation principles in plumbing licensing examinations.

1-1 What is the BOCA?

BOCA is an acronym for Building Officials and Code Administration or the *Basic Plumbing Code*.

1-2 What are the seven classifications of pipe materials contained in the BOCA code?

The seven classifications are as follows:

- **Water service pipe.** This type of pipe may be made of asbestos cement, brass, cast iron, copper, plastic, galvanized iron, or steel.
- **Water distribution pipe.** This type of pipe may be made of brass, copper, plastic, or galvanized steel.
- **Aboveground drainage and vent pipe.** This type of pipe may be made of brass, cast iron, copper (type K, L, M, or DWV), galvanized steel, or plastic.
- **Underground drainage and vent pipe.** This type of pipe can be made of brass, cast iron, copper (type K, L, M, or DWV), galvanized steel, or plastic.
- **Building sewer pipe.** This type of pipe may be made of asbestos cement, bituminized fiber, cast iron, copper (type K or L), concrete, plastic, or vitrified clay.
- **Building storm sewer pipe.** This type of pipe may be made of asbestos cement, bituminized fiber, cast iron, concrete, copper (type K, L, M, or DWV), or vitrified tile.
- **Subsoil drainpipe.** This type of pipe may be made of asbestos cement, bituminized fiber, cast iron, plastic, styrene rubber, or vitrified clay.

1-3 What are the six types of corrosion of which a plumber should have knowledge?

The six types of corrosion are as follows:

- **Concentrated cell corrosion.** This is an electrical effect caused by differences in composition of a solution in contact with a single metal or alloy.
- **Dezincification corrosion.** This form of corrosion usually occurs with alloys of copper and zinc in which the zinc content is more than 15 percent. The zinc is chemically leached out of the copper-zinc compound.
- **Graphite-type corrosion.** This type of corrosion is commonly related to cast iron and resembles zincification. It means that

the iron in the gray cast iron may be leached out. This leaves behind a porous layer of graphite over the remaining unattacked iron.

- **Stress-accelerated corrosion.** Under high stress, some of the yellow brass alloys may have a quantity of metal dissolve and create a mechanical weakening in the material.
- **Galvanic corrosion.** This corrosion is created by various sources creating electrochemical rather than electrical differences. This occurs when there are two dissimilar metals making contact with a solution *covering* the area.
- **Pitting and local corrosion.** This type of corrosion usually occurs when there is a breakdown in the passive films that a metal or alloy builds up in a corroding solution. This may also occur as a result of accelerated corrosion. This may have a surface blemish or nonmetallic inclusion.

Multiple-Choice Exercises

Select the right answer and blacken *a*, *b*, *c*, or *d* on your practice answer sheet. (See Appendix C for answers.)

1. Water service pipe is classified into seven groups. Which of the following is not one of them?
   a. Asbestos cement
   b. Copper
   c. Plastic
   d. Lead

2. Water distribution pipe is classified into four categories. Which of the following is *not* one of them?
   a. Lead
   b. Copper
   c. Plastic
   d. Brass

3. What is the BOCA?
   a. British Overseas Aircraft Company
   b. Broadband on all cars
   c. Building Officials and Code Administration
   d. Building Offices and Code Alterations

4. The drainage of sewage and wastes from a plumbing system shall not be discharged into subsurface soil or into a water surface unless which of the following occurs?

a. It has been properly treated.

b. It meets the Clean Water Act's guidelines.

c. Environmental officials approve.

d. It is cleaner than when it was first used.

5. Which of the following is a type of corrosion that a plumber should know about?

a. Concentrated cell corrosion

b. Acid corrosion

c. Degenerative corrosion

d. Just plain corrosion

6. Copper pipe is classified into four categories. Which of these is not one of them?

a. K

b. L

c. DWV

d. N

7. Underground drainage and vent pipe can be made of which of the following?

a. Brass

b. Cast iron

c. Concrete

d. Plastic

8. Buildings with plumbing fixtures and devices shall be provided with a supply of water in sufficient volume and _____ to enable them to operate in a satisfactory manner at all times.

a. pressure

b. quantity

c. quality

d. velocity

9. Galvanic action/galvanic corrosion is caused by which of the following?

a. The galvanizing action during the application of a zinc coating of steel.

b. The plating action of unlike metals.

c. Two dissimilar metals coming in contact with a liquid or electrolyte covering them.

d. Two acids and galvanic steel being mixed.

10. Sewage or other wastes from a plumbing system shall not be discharged into subsurface soil or into a water _____ unless it has first been properly treated.

a. surface

b. hole

c. pond

d. river

11. Buildings with plumbing fixtures and devices shall be provided with a supply of water in sufficient volume and _____ to enable them to operate in a satisfactory manner at all times.

a. pressure

b. quantity

c. quality

d. velocity

Chapter 2

Plumbing Drawings, Material Takeoff Procedures, and Typical Plumbing System Layout

Technical drawings and specifications are the graphic language of the engineer. Definite rules of usage provide identical meanings in most localities. Observance of these rules allows people anywhere to read engineering drawings and specifications.

Specifications and drawings provide information concerning size and shape so that the craftsperson can use building materials to create the finished product desired by the planner.

A basic understanding of plumbing symbols and terminology is necessary so that an applicant for a plumber's license can easily understand the printed language used in plumber's license examination questions.

By definition, all piping, apparatus, and fixtures for water distribution and waste disposal systems in a building are termed *plumbing*. Pipe, fittings, unions, and valves compose the piping system.

Pipe

Pipe is supplied in straight lengths in sections from 12 feet to 20 feet long. Standard piping of wrought iron or steel up to 12 inches in diameter is classified by its nominal inside diameter. The actual inside diameter may vary for a given nominal size (such as standard pipe, heavy pipe, or extra heavy pipe), depending on pipe weight factors. The external diameter is normally the same for all three weights. Pipe above 12 inches is classified by its actual diameter.

Brass and copper piping are classified by the same nominal sizes as iron pipe with two weights for each size—extra strong and regular.

Fittings

Various lengths of pipe are connected to create a desired length or form by fittings that also provide continuity and direction changes. Pipe continuity is provided by the use of couplings, nipples, and reducers. Tees, crosses, and elbows are used to change direction of a run of piping. A *cap* is used to close an open pipe, a *plug* is used to close an open fitting, and a *bushing* is used to reduce the size of an opening.

Unions provide convenient connections that can be easily unmade. Screwed unions are of three-piece design. Two pieces are screwed to the ends of the pipes being connected, and the third draws them together by screwing onto the first piece and bearing against the shoulder of the second. Flow in a piping system is regulated by valves that are specified by type, material, size, and working pressures.

Water Distribution Systems

Basically, the water distribution system of a building is the inside water supply plumbing system. Water requirements to each fixture are supplied from the water main with reduced-diameter pipe connections.

Two types of water supply systems are in general use: the downfeed and the upfeed. In a *downfeed system,* the supply tank is located on the roof or high point on the inside of a building. Water flows by gravity to the fixtures after being mechanically pumped to the supply tank.

Upfeed systems are more common. Water is supplied from an elevated storage tank or water tower. The lowest elevation point of the water tank is so designed and located that it is above the highest point of the water distribution system. Water fills the supply system lines as it seeks its water level.

Materials for hot and cold water supply systems are generally of copper, brass, wrought iron, or galvanized steel. Fittings are made of the same materials as the pipe. Material types are designated by job specifications and size and are noted on the drawings or by special instructions.

Vents

The following key points should be kept in mind regarding vents:

- **Soil pipes and soil vents.** *Soil pipe* can be defined as that part of the system that carries away the solid and liquid wastes from the water closet or similar fixtures. A *soil vent* is the vent pipe that extends through the roof and relieves the unequal air pressure resulting from the waste falling or running down the soil pipe. In addition, a vent prevents traps from siphoning themselves dry by introducing air, thus breaking the suction resulting from the falling water.

- **Waste pipe.** *Waste pipe* can be classified as the piping in the plumbing system that carries all other wastes except those from the toilet. Waste vents extend through the roof or into the soil vent above the highest fixture locations. Vents introduce air into the plumbing system and break the suction resulting from the falling or draining of waste pipe water.

* **Main vents and revents.** *Main vents* serve the major bathroom group in the house. *Revents* are secondary vents used to vent fixtures more than 8 feet from the main vent and to relieve air pressures on fixtures located vertically downstream from the water closet.

Plumbing Drawings

For buildings of average or normal size, plumbing drawings are usually designed in combination with other utility requirements. This assumes that the various tradespeople can readily identify symbols or instructions pertaining to their trades and, from these, proceed with their work requirements. Plumbing drawings for large jobs are placed on separate individual plumbing drawing sheets and identified as plumbing drawings. The following key points should be kept in mind regarding plumbing drawings:

* **Plumbing fixtures on drawings.** Fixtures on drawings are represented symbolically and drawn to scale in planned locations. Dimensions are generally omitted. Many symbols are drawn touching wall lines. Examples of these are wall lavatories, corner bath, and low-tank water closets. Figure 2-1 illustrates typical standard plumbing symbols.
* **Piping on drawings.** Piping is generally shown on drawings by a single line. Valve and fitting symbols are used only for supply piping and are not used for waste lines. Specifications designate joint types. Exact pipe length runs are in the field at the job site. A single-line pipe symbol on the plans between fixtures is for the purpose of indicating pipe location. Pipe runs are identified for nominal size by placing the printed figure size next to the pipe with an arrowhead extending to the pipe. A new dimension is given whenever a pipe changes size.
* **Cold-water piping.** Cold-water supply pipes are identified by a heavy dashed line consisting of alternating long and short dashes, as illustrated in Figure 2-2.
* **Hot-water piping.** Hot-water supply pipes are shown on drawings by a single heavy dashed line consisting of a series of long dashes or a series of short dashes. They are labeled as indicated in Figure 2-2.
* **Drainage lines.** Drainage or waste pipe is indicated on drawings by a heavy solid line. Figure 2-2 shows a drainage line drawing symbol.

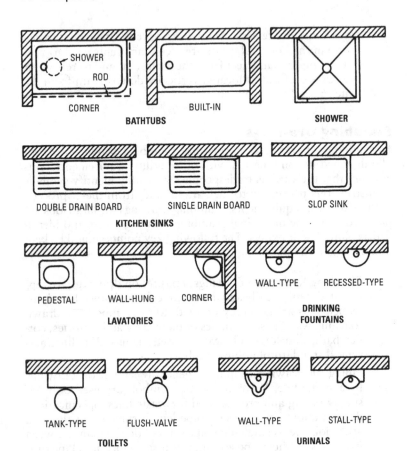

Figure 2-1 Typical plumbing fixtures symbols.

SYMBOL	PLAN	INITIALS	ITEM
————————	O	D.	DRAINAGE LINE
—•—•—•—	O	V.S.	VENT STACK
– – – – –	o	C.W.	COLD WATER LINE
— – — – —	o	H.W.	HOT WATER LINE

Figure 2-2 Typical plumbing drawing symbols.

- **Vent stacks.** On plumbing drawings, vertical vent stacks appear in section and are shown as circles drawn to scale. They may also appear as heavily broken, evenly spaced solid lines as shown in Figure 2-2.
- **Drawing legends.** Plumbing instructions or symbols are generally organized under the heading "Legend" and placed in a convenient location on the plumbing drawing.
- **Vertical views.** Vertical piping views are used only to clarify complicated piping systems. Sections and elevations are then shown on the drawing.
- **Riser diagrams.** Riser diagrams are elevations or perspective views showing stacks and risers, as illustrated in Figure 2-3. Generally, riser diagrams are shown only for complicated jobs.
- **Drawing installation details.** Plumbing drawing installation details are not generally placed on the plumbing drawings. Specifications indicate type and quality of fixtures, piping, and fittings. Tradespeople will install materials in accordance with standard trade practices, drawing and specification instructions, and local plumbing code instructional requirements.

Plumbing Takeoff Procedures

Plumbing takeoff requires a study of the drawings and specifications and the compiling and listing of all the materials including fixtures and appurtenances necessary to complete a specific job.

In the process, the estimator or person compiling the material list must consider the piping waste or loss factor. Extra pipe quantities must be added to the list to offset losses caused by cutting and fitting.

- **Fixtures.** Fixture takeoff requirements are determined after a study of the drawings and specifications. Requirements are generally listed in the specifications or on the drawings by name, stock number, and required quantity.
- **Pipe.** On the takeoff sheet, piping is listed by size, material type, stock number (if available), and lineal feet required. Following is a partial example list:

Size/Material Type	Stock Number	Lineal Feet
¹/₂-inch copper tubing	Stock number	40
¹/₂-inch galv. steel pipe	Stock number	300
³/₄-inch galv. steel pipe	Stock number	80

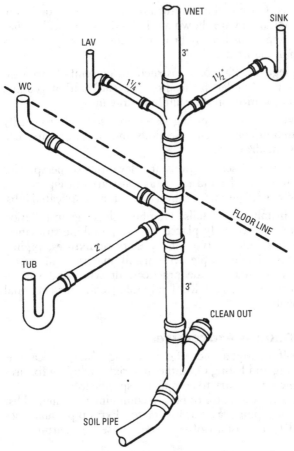

Figure 2-3   A riser diagram showing stack venting for one bath group.

- **Fittings.** Fittings are listed on the takeoff or estimate sheet by size, material, classification, stock number, and quantity requirements. A partial typical listing follows:

Size/Material Type	Stock Number	Quantity
$1/2$-inch galv. couplings	Stock number	12 each
$1/2$-inch galv. cast-iron elbows, 90°	Stock number	30 each
$3/4$-inch galv. cast-iron elbows, 45°	Stock number	15 each

* **Soil pipe.** Soil pipe is commonly listed by standard length requirements, that is, by size, material, and strength or weight. Standard shipment is in 5-foot lengths. Fittings are classified by size, material, classification, weight or strength, and quantity.

In soil pipe, a 90° elbow is referred to as a quarter-bend and a 45° elbow as an eighth-bend. Material is listed as follows:

Size/Material	Weight/Strength/ Quantity
4-inch extra-heavy cast-iron soil pipe (5-foot lengths)	10 lengths
4-inch eighth-bend extra-heavy cast iron	2 each
4-inch quarter-bend extra-heavy cast iron	4 each

Typical Complete Home Plumbing System

Two separate systems constitute the home plumbing system (see Figure 2-4): water supply and water disposal. Water in the water supply system is under about 50 pounds of pressure per square inch (psi). Water supply pipes can be somewhat small in diameter but still carry enough water. Water in the disposal—called the drain-waste-vent (DWV) system—flows by gravity. Therefore, its piping and fittings are larger in diameter in order to carry the required flow without clogging or backing up. Both systems operate safely if designed properly.

Water supply and DWV systems are never connected to each other. The entry of contaminated water into drinking water can lead to serious illness or even death.

Multiple-Choice Exercises

Select the right answer and blacken *a*, *b*, *c*, or *d* on your practice answer sheet. (See Appendix C for answers.)

1. Pipe is supplied in straight lengths in sections from 12 feet to _____ feet long.

   a. 40

   b. 20

   c. 30

   d. 50

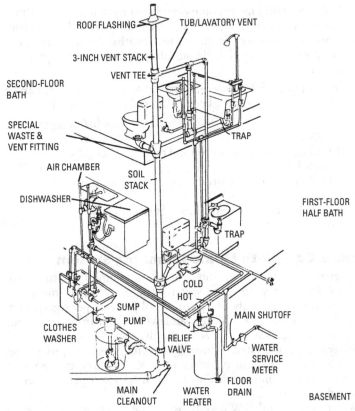

ROOF FLASHING

TUB/LAVATORY VENT

3-INCH VENT STACK

VENT TEE

SECOND-FLOOR
BATH

SPECIAL
WASTE &
VENT FITTING

TRAP

AIR CHAMBER

SOIL
STACK

DISHWASHER

FIRST-FLOOR
HALF BATH

TRAP

COLD
HOT

SUMP
PUMP

MAIN SHUTOFF

CLOTHES
WASHER

RELIEF
VALVE

WATER
SERVICE
METER

FLOOR
DRAIN

MAIN
CLEANOUT

WATER
HEATER

BASEMENT

Figure 2-4   A typical complete home plumbing system. *(Courtesy Genova Inc.)*

2. Flow in a piping system is regulated by ___ that are specified by type, material, size, and working pressures.

   a. relays

   b. valves

   c. check valves

   d. computers

3. Two types of water supply systems are in general use. They are the downfeed and the ___.

   a. gravity feed

   b. upfeed

c. pumped feed

d. normal pressure feed

4. A soil vent pipe is the pipe that extends through the roof and relieves the _____ air pressure resulting from the waste falling or running down the soil pipe.

a. high

b. low

c. equal

d. unequal

5. Revents are _____ vents used to vent fixtures more than 8 feet from the main vent and to relieve air pressures on fixtures located vertically downstream from the water closet.

a. primary

b. secondary

c. tertiary

d. open

6. Riser diagrams are elevations or perspective views showing stacks and risers. They generally are shown only for _____ jobs.

a. noncomplicated

b. complicated

c. simple

d. easy

7. Fixture takeoff requirements are determined after a study of the _____ and specifications.

a. floor plan

b. terrain layout

c. city sewer system

d. drawings

8. Soil pipe is commonly listed by standard length requirements, size, the material, and the strength or _____.

a. weakness

b. weight

c. diameter

d. composition

9. Various lengths of pipe are often connected to form a desired length. They may also form, by means of fittings, _____ and direction changes.

   a. continuity

   b. height

   c. length

   d. connections

10. Materials usually used for hot and cold water supply systems are not made of _____.

    a. copper

    b. brass

    c. wrought iron

    d. asbestos cement

Essay Question Exercise

These *essay-type questions* require answers that a first reading of the material will most likely not furnish. It is suggested that you read the material for the chapter and then try to answer all the multiple-choice questions you have answers for. Then, take a new sheet of paper and answer the questions here by using the book. This way you will, at the very least, review the material and keep the proper answers in your mind for the final test when you apply for your license.

1. Why are the specifications and drawings for a plumbing system so important?

2. Describe a typical water distribution system.

3. Describe how vents, pipe, and fittings are used in a distribution system.

4. What are some typical plumbing fixture symbols found on a drawing?

5. What is meant by plumbing takeoff?

6. What does DWV stand for?

Chapter 3

Mathematics

A working knowledge of mathematics is important and beneficial to all trades people, particularly plumbers. Therefore, a general review of mathematics, with some reference tables, follows.

Common Symbols

For convenience and brevity, the various processes in mathematics are usually indicated by symbols. Table 3-1 shows some of the commonly used symbols.

Table 3-1 Common Symbols

Symbol	Description
=	Equal to, or equality.
−	Minus, less, or subtraction.
+	Plus, or addition.
×	Multiplied by, or multiplication.
÷ or /	Divided by, or division.
2,3	Indexes or powers. The number to which these superscripts are added should be squared or cubed. Thus, 2^2 means "2 squared," while 2^3 means "2 cubed."
$\sqrt{}$	Radical sign. This means that the square root is to be extracted from the number before which it is placed.
$\sqrt[3]{}$	Radical sign. This means that the cube root is to be extracted from the number before which it is placed.
.	The decimal point, which means decimal parts. Thus, 2.5 means $2^5/_{10}$; 0.46 means $^{46}/_{100}$.
°	Degrees(°Fahrenheit or °Celsius).
π	Ratio of the circumference of a circle to its diameter, numerically 3.1415927.

Abbreviations

In addition to these symbols, a plumber should be familiar with the following abbreviations.

Common Abbreviations

Table 3-2 shows common abbreviations.

Special Abbreviations

Table 3-3 shows special abbreviations.

Table 3-2 Common Abbreviations

Abbreviation	Description
A or a	Area
A.W.G.	American wire gage
bbl	Barrels
B or b	Breadth
bhp	Brake horsepower
B.M.	Board measure
Btu	British thermal units
B.W.G.	Birmingham wire gage
C of g	Center of gravity
cu.	Cubic
cu. in	Cubic inches
cyl.	Cylinder
D or d	Depth or diameter
deg.	Degrees
diam.	Diameter
F	Coefficient of friction; Fahrenheit
ft-lbs	Foot pounds
gals.	Gallons
gpm	Gallons per minute
H or h	Height, or head of water
hp	Horsepower
ihp	Indicated horsepower
lbs	Pounds
lbs per sq. in	Pounds per square inch (psi)
o.d.	Outside diameter (pipes)
psi	Pounds per square inch
R or r	Radius
rpm	Revolution per minute
$\square'$	Square feet
sq. ft	Square foot
sq. in	Square inch
$\square''$	Square inches
sq. yd	Square yard
temp.	Temperature
V or v	Velocity
vol.	Volume
W.I.	Wrought iron

Table 3-3   Special Abbreviations

Abbreviation	Description
atm. pres.	Standard atmospheric pressure
av.	Average
cal. val.	Calorific value
C_p	Specific heat at constant pressure
C_v	Specific heat at constant volume
cm	Centimeters
evap.	Evaporation
g.	Gravity acceleration
Hg	Mercury
in. Hg	Inches of mercury (pressure)
kg	Kilograms
log	Logarithm to the base of 10
$\log_e$	Logarithm to the base "e"
min.	Minute
mol. wt.	Molecular weight
sat	Dry saturated (steam)
sec	Second
sup	Superheated

Mathematical Definitions

Table 3-4 shows common mathematical terms and corresponding definitions.

Table 3-4   Mathematical Definitions

Term	Definition
Abstract number	A number that does not refer to any particular object.
Acute triangle	A triangle with three acute angles.
Altitude (of a prism)	The perpendicular distance between its bases.
Altitude (of a pyramid or cone)	The perpendicular distance from its vertex to the plane of its base.
Analysis	The process of investigating principles and solving problems independently of set rules.
Angle	The difference in direction of two lines proceeding from the same point called the vertex.

(*continued*)

Table 3-4 (continued)

Term	Definition
Area	The surface included within the lines that bound a plane figure.
Arithmetic	The science of numbers and the art of computation.
Base (of a triangle)	The side on which it may be supposed to stand.
Circle	A plane figure bounded by a curved line (called the *circumference*), every point of which is equally distant from a point within (called the *center*).
Complex fraction	A fraction whose numerator or denominator is called a fraction.
Compound fraction	A fraction of a fraction.
Concrete number	A number used to designate objects or quantities.
Cone	A body having a circular base, and whose convex surface tapers uniformly to the vertex.
Cubic measure	A measure of volume involving three dimensions, namely, length, breadth, and thickness.
Decimal scale	A scale in which the order of progression is uniformly ten.
Demonstration	The process of reasoning by which a truth or principle is established.
Denomination	The name of the unit of a concrete number.
Denominator	The part of a fraction that is below the line signifying division.
Diameter (of a circle)	A line passing through its center and terminated at both ends by the circumference.
Diameter (of a sphere)	A straight line passing through the center of a sphere, and which is terminated at both ends by its surface.
Even number	A number that can be exactly divided by two.
Factors	One of two or more quantities that, when multiplied together, produce a given quantity.
Factors of a number	Factors are numbers that, when multiplied together, make that number.

(*continued*)

Table 3-4 (continued)

Term	Definition
Fraction	A number that expresses parts of a whole thing or quantity.
Geometry	That branch of pure mathematics that deals with the measurements, properties, and relationships of points, lines, angles, surfaces, and solids.
Greatest common divisor	The greatest number that will exactly divide two or more numbers.
Hypotenuse (of a right triangle)	The side opposite the right angle.
Improper fraction	A fraction whose numerator equals or exceeds its denominator.
Integer	A number that represents whole things.
Least common multiple	The least number that is exactly divisible by two or more numbers.
Mathematics	The science of quantity.
Measure	That by which the extent, quantity, capacity, volume, or dimensions in general are ascertained by some fixed standard.
Mensuration	The process of measuring.
Multiple of a number	Any number exactly divisible by that number.
Number	A unit or collection of units.
Numerator	The part of a fraction that is above the line and signifies the number of parts of the denominator taken.
Percentage	Rate per hundred.
Perpendicular (of a right triangle)	The side that forms a right angle with the base.
Power	The product arising from multiplying a number.
Proper fraction	A fraction whose numerator is less than its denominator.
Quantity	That which can be increased, diminished, or measured.
Radius (of a circle)	A line extending from its center to any point on the circumference; it is one half the diameter.
Radius (of a sphere)	A straight line drawn from the center to any point on the surface.

(continued)

Table 3-4 (continued)

Term	Definition
Right triangle	A triangle that has a right angle.
Root	A factor repeated to produce a power.
Rule	A prescribed method of performing an operation.
Scale	The order of progression on which any system of notation is founded.
Simple fraction	A fraction whose numerator and denominator are whole numbers.
Simple number	An abstract or concrete number of but one denomination.
Slant height (of a cone)	A straight line from the vertex to the circumference of the base.
Slant height (of a pyramid)	The perpendicular distance from its vertex to one of the sides of the base.
Sphere	A body bounded by a uniformly curved surface, all the points of which are equally distant from a point within called the center.
Square	A rectangle whose sides are equal.
Triangle	A plane figure bounded by three sides, and having three angles.
Uniform scale	A scale in which the order of progression is the same throughout the entire succession of units.

Notation and Numeration

By definition, *notation* in arithmetic is the writing down of figures to express a number. A *numeration* is the reading of the number or collection of figures already written.

Elementary Operations

The following 10 formulas include the elementary operations of arithmetic:

- The sum equals all the parts added.
- The difference equals the minuend minus the subtrahend.
- The minuend equals the subtrahend plus the difference.
- The subtrahend equals the minuend minus the difference.
- The product equals the multiplicand times the multiplier.

- The multiplicand equals the product divided by the multiplier.
- The multiplier equals the product divided by the multiplicand.
- The quotient equals the dividend divided by the divisor.
- The dividend equals the quotient times the divisor.
- The divisor equals the dividend divided by the quotient.

Fractions

If a unit or whole number is divided into two equal parts, one of these parts is called one-half, written $\frac{1}{2}$.

To reduce a common fraction to its lowest terms:

Rule: Divide both terms by their greatest common divisor.

Example:

$$\frac{9}{15} = \frac{(9 \div 3)}{(15 \div 3)} = \frac{3}{5}$$

To change an improper fraction to a mixed number:

Rule: Divide the numerator by the denominator. The quotient is the whole number, and the remainder placed over the denominator is the fraction.

Example:

$$\frac{23}{4} = 5\frac{3}{4}$$

To change a mixed number to an improper fraction:

Rule: Multiply the whole number by the denominator of the fraction. Add the numerator to the product and place the sum over the denominator.

Example:

$$1\frac{3}{8} = \frac{(8 \times 1) + 3}{8} = \frac{11}{8}$$

To reduce a compound to a simple fraction, and to multiply fractions:

Rule: Multiply the numerators together for a new numerator and the denominators together for a new denominator.

Example:

$$\frac{1}{2} \text{ of } \frac{2}{3} = \frac{2}{6} \text{ or } \frac{1}{2} \times \frac{2}{3} = \frac{1 \times 2}{2 \times 3} = \frac{2}{6}$$

To reduce a complex fraction to a simple fraction:

Rule: The numerator and denominator must first be given the form of a simple fraction; then multiply the numerator of the upper fraction by the denominator of the lower for the new numerator, and the denominator of the upper by the numerator of the lower for the new denominator.

Example:

$$\frac{\frac{7}{8}}{1\frac{3}{4}} = \frac{\frac{7}{8}}{\frac{7}{4}} = \frac{7 \times 4}{8 \times 7} = \frac{28}{56} = \frac{1}{2}$$

To add fractions:

Rule: Reduce them to a common denominator, add the numerators, and place their sum over the common denominator.

Example:

$$\frac{1}{2} + \frac{1}{4} = \frac{4+2}{8} = \frac{6}{8} = \frac{3}{4}$$

To subtract fractions:

Rule: Reduce them to a common denominator, subtract the numerators, and place the difference over the common denominator.

Example:

$$\frac{1}{2} - \frac{1}{4} = \frac{4-2}{8} = \frac{2}{8} = \frac{1}{4}$$

To multiply fractions:

Rule (for multiplying by a whole number): Multiply the numerator or divide the denominator by the whole number.

Example:

$$\frac{1}{2} \times 3 = \frac{3}{2} = 1\frac{1}{2}$$

To divide fractions:

Rule (for dividing by a whole number): Divide the numerator or multiply the denominator by the whole number.

Example:

$$(\text{dividing})\frac{10}{13} \div 5 = \frac{2}{13}$$

$$(\text{multiplying})\frac{10}{13} \times \frac{1}{5} = \frac{10}{65} = \frac{2}{13}$$

Rule (for dividing by a fraction): Invert the divisor and proceed as in multiplication.

Example:

$$\frac{3}{4} \div \frac{5}{7} = \frac{3}{4} \times \frac{7}{5} = \frac{21}{20} = 1\frac{1}{20}$$

Powers of Numbers

The square of a number is its second power; the cube is its third power. Thus, the square of $2 = 2 \times 2 = 4$; the cube of $2 = 2 \times 2 \times 2 = 8$.

The power to which a number is raised is indicated by a small superior figure called an *exponent*. Thus:

$$2^2 = 2 \times 2 = 4; \quad 2^3 = 2 \times 2 \times 2 = 8$$

Roots of Numbers (Evolution)

In the equation $2 \times 2 = 4$, the number 2 is the root for which the power (4) is produced. The radical sign $\sqrt{}$ placed over a number means the root of the number is to be extracted. Thus, $\sqrt{4}$ means that the square root of 4 is to be extracted. The index of the root is a small figure placed over the radical. For example, $\sqrt[3]{27}$ means that the cube root of 27 is to be extracted.

Rule (for square roots): As shown in the following example, point off the given number into groups of two places each, beginning with units. If there are decimals, point these off likewise, beginning at the decimal point and supplying as many ciphers as may be needed. Find the greatest number whose square is less than the first left-hand group and place it as the first figure in the quotient. Subtract its square from the left-hand group and annex the two figures of the second group to the remainder for a dividend. Double the first figure of the quotient for a partial divisor; find the number of times the latter is contained in the dividend, exclusive of the right-hand figure in the quotient, and annex it to the right of the partial divisor, thus forming the complete divisor. Multiply this divisor by the second figure in the quotient and subtract the product from the dividend. To the remainder, bring down the next group and proceed as before, in each case doubling the figures in the root already found to obtain the trial divisor. Should the product of the second figure in the root by the completed divisor be greater than the dividend, erase the second figure from both the quotient and the divisor and substitute the next smaller figure, or one small enough to make the product of the second figure by the divisor less than or equal to the dividend.

Example:

$$\sqrt{3.14'15'92'65'36} \quad \lfloor 1.77245 +$$

$$
\begin{array}{r|l}
 & 1 \\
\hline
27 & 214 \\
 & 189 \\
\hline
347 & 2515 \\
 & 2429 \\
\hline
3542 & 8692 \\
 & 7084 \\
\hline
35444 & 160865 \\
 & 141776 \\
\hline
354485 & 1908936 \\
 & 1772425 \\
\hline
 & 136511
\end{array}
$$

Rule (for cube roots): As shown in the following example, separate the number into groups of three figures each, beginning at the units. Find the greatest cube in the left-hand group and write its root for the first figure of the required root. Cube this root, subtract the result from the left-hand group, and to the remainder annex the next group for a dividend. For a partial divisor, take three times the square of the root already found (considered as hundreds) and divide the dividend by it. The quotient (or the quotient diminished) will be the second figure of the root. To this partial divisor add three times the product of the first figure of the root (considered as tens) multiplied by the second figure and add the square of the second figure of the root. This sum will be the complete divisor. Multiply the complete divisor by the second figure of the root, subtract the product from the dividend, and to the remainder annex the next group for a new dividend. Proceed in this manner until all groups have been annexed. The result will be the cube root required. Table 3-5 can be a great help in determining the square, cube, square root, or cube root of numbers up to 100.

Example:

$$\sqrt[3]{1{,}881{,}365{,}963{,}625} \quad \lfloor 12345$$

300×1^2	$= 300$	881
30×1	$\times 2 = 60$	
	$2^2 = 4$	
	364	728
		153365
300×12^2	$= 43200$	
30×12	$\times 3 = 1080$	
	$3^2 = 9$	
	44289	132867
		20498963
300×123^2	$= 4538700$	
30×123	$\times 4 = 14760$	
	$4^2 = 16$	
	4553476	18213904
		2285059625
300×1234^2	$= 456826800$	
$30 \times 1234 \ \times 5 =$	185100	
	$5^2 = 25$	
	457011925	2285059625

Table 3-5   Squares, Cubes, Square Roots, and Cube Roots

No.	Square	Cube	Square Root	Cube Root	Reciprocal
1	1	1	1.00000	1.00000	1.00000
2	4	8	1.41421	1.25992	0.50000
3	9	27	1.73205	1.44224	0.33333
4	16	64	2.00000	1.58740	0.25000
5	25	125	2.23606	1.70997	0.20000
6	36	216	2.44948	1.81712	0.16666
7	49	343	2.64575	1.91293	0.14285
8	64	512	2.82842	2.00000	0.12500
9	81	729	3.00000	2.08008	0.11111
10	100	1000	3.16227	2.15443	0.10000

(continued)

Table 3-5 (continued)

No.	Square	Cube	Square Root	Cube Root	Reciprocal
11	121	1331	3.31662	2.22398	0.09090
12	144	1728	3.46410	2.28942	0.08333
13	169	2197	3.60555	2.35133	0.07602
14	196	2744	3.74165	2.41014	0.07142
15	225	3375	3.87298	2.46621	0.06666
16	256	4096	4.00000	2.51984	0.06250
17	289	4913	4.12310	2.57128	0.05882
18	324	5832	4.24264	2.62074	0.05555
19	361	6859	4.35889	2.66840	0.05263
20	400	8000	4.47213	2.71441	0.05000
21	441	9621	4.58257	2.75892	0.04761
22	484	10,648	4.69041	2.80203	0.04545
23	529	12,167	4.79583	2.84386	0.04347
24	576	13,824	4.89897	2.88449	0.04166
25	625	15,625	5.00000	2.92401	0.04000
26	676	17,576	5.09901	2.96249	0.03846
27	729	19,683	5.19615	3.00000	0.03703
28	784	21,952	5.29150	3.03658	0.03571
29	841	24,389	5.38516	3.07231	0.03448
30	900	27,000	5.47722	3.10723	0.03333
31	961	29,791	5.56776	3.14138	0.03225
32	1024	32,768	5.65685	3.17480	0.03125
33	1089	35,937	5.74456	3.20753	0.03030
34	1156	39,304	5.83095	3.23961	0.02941
35	1225	42,875	5.91607	3.27106	0.02857
36	1296	46,656	6.00000	3.30192	0.02777
37	1369	50,653	6.08276	3.33222	0.02702
38	1444	54,872	6.16441	3.36197	0.02631
39	1521	59,319	6.24499	3.39121	0.02564
40	1600	64,000	6.32455	3.41995	0.02500
41	1681	68,921	6.40312	3.44821	0.02439
42	1764	74,088	6.48074	3.47602	0.02380
43	1849	79,507	6.55743	3.50339	0.02325
44	1936	85,184	6.63324	3.53034	0.02272
45	2025	91,125	6.70820	3.55689	0.02222
46	2116	97,336	6.78233	3.58304	0.02173

(continued)

Table 3-5 *(continued)*

No.	Square	Cube	Square Root	Cube Root	Reciprocal
47	2209	103,823	6.85565	3.60882	0.02127
48	2304	110,592	6.92820	3.63424	0.02083
49	2401	117,649	7.00000	3.65930	0.02040
50	2500	125,000	7.07106	3.68403	0.02000
51	2601	132,651	7.14142	3.70842	0.01960
52	2704	140,608	7.21110	3.73251	0.01923
53	2809	148,877	7.28010	3.75628	0.01886
54	2916	157,464	7.34846	3.77976	0.01851
55	3025	166,375	7.41619	3.80295	0.01818
56	3136	175,616	7.48331	3.82586	0.01785
57	3249	185,193	7.54983	3.84850	0.01754
58	3364	195,112	7.61577	3.87087	0.01724
59	3481	205,379	7.68114	3.89299	0.01694
60	3600	216,000	7.74596	3.91486	0.01666
61	3721	226,981	7.81024	3.93649	0.01639
62	3844	238,328	7.87400	3.95789	0.01612
63	3969	250,047	7.93725	3.97905	0.01587
64	4096	262,144	8.00000	4.00000	0.01562
65	4225	274,625	8.06225	4.02072	0.01538
66	4356	287,496	8.12403	4.04124	0.01515
67	4489	300,763	8.18535	4.06154	0.01492
68	4624	314,432	8.24621	4.08165	0.01470
69	4761	328,500	8.30662	4.10156	0.01449
70	4900	343,000	8.36660	4.12128	0.01428
71	5041	357,911	8.42614	4.14081	0.01408
72	5184	373,248	8.48528	4.16016	0.01388
73	5329	389,017	8.54400	4.17933	0.01369
74	5476	405,224	8.60232	4.19833	0.01351
75	5625	421,875	8.66025	4.21716	0.01333
76	5776	438,976	8.71779	4.23582	0.01315
77	5929	456,533	8.77496	4.25432	0.01298
78	6084	474,552	8.83176	4.27265	0.01282
79	6241	493,039	8.88819	4.29084	0.01265
80	6400	512,000	8.94427	4.30886	0.01250
81	6561	531,441	9.00000	4.32674	0.01234
82	6724	551,368	9.05538	4.34448	0.01219

(continued)

Table 3-5 (continued)

No.	Square	Cube	Square Root	Cube Root	Reciprocal
83	6889	571,787	9.11043	4.36207	0.01204
84	7056	592,704	9.16515	4.37951	0.01190
85	7225	614,125	9.21954	4.39682	0.01176
86	7396	636,056	9.27361	4.41400	0.01162
87	7569	658,503	9.32737	4.43104	0.01149
88	7744	681,472	9.38083	4.44796	0.01136
89	7921	704,969	9.43398	4.46474	0.01123
90	8100	729,000	9.48683	4.48140	0.01111
91	8281	753,571	9.53939	4.49794	0.01098
92	8464	778,688	9.59166	4.51435	0.01086
93	8649	804,357	9.64365	4.53065	0.01075
94	8836	830,584	9.69535	4.54683	0.01063
95	9025	857,375	9.74679	4.56290	0.01052
96	9216	884,736	9.79795	4.57885	0.01041
97	9409	912,673	9.84885	4.59470	0.01030
98	9604	941,192	9.89949	4.61043	0.01020
99	9801	970,299	9.94987	4.62606	0.01010
100	10,000	1,000,000	10.00000	4.64158	0.01000

The Metric System

The important feature of the metric system is that it is based on the decimal scale. The metric system is the decimal system of measures and weights, with the meter and the gram as the bases. The unit of length (the meter) was intended to be (and is very nearly) one ten-millionth part of the distance measured on a meridian from the equator to the pole, or 39.37079 inches. The other primary units of measure (such as the square meter, the cubic meter, the liter, and the gram) are based on the meter.

The following lists give the metric system of weights and measures.

Milli expresses the 1000th part

Centi expresses the 100th part

Deci expresses the 10th part

Deca expresses 10 times the value

Hecto expresses 100 times the value

Kilo expresses 1000 times the value

Length
Following are some common metric measurements of length:

1 mm = 1 millimeter (mm) = $\frac{1}{1000}$ of a meter = 0.03937 in
10 mm = 1 centimeter (cm) = $\frac{1}{100}$ of a meter = 0.3937 in
10 cm = 1 decimeter (dm) = $\frac{1}{10}$ of a meter = 3.937 in
10 dm = 1 meter (m) = 1 meter = 39.3707 in. = 3.28 ft
10 m = 1 decameter (da) = 10 meters = 32.809 ft
10 da = 1 hectometer (hm) = 100 meters = 328.09 ft
10 hm = 1 kilometer (km) = 1000 meters = 0.62137 mile
10 km = 1 myriameter = 10,000 meters = 6.2138 mile

Square Measure
Following are some common metric measurements of square measure:

1 sq. centimeter (cm^2) = 0.1550 sq. in
1 sq. decimeter (dm^2) = 15.5 sq. in
1 sq. meter (m^2) = 1.196 sq. yd
1 are = 3.954 sq. rd
1 hectare = 2.47 acres
1 sq. kilometer (km^2) = 0.386 sq. mile
1 sq. in = 6.452 sq. centimeters (cm^2)
1 sq. ft = 9.2903 sq. decimeters (dm^2)
1 sq. yd = 0.8361 sq. meter (m^2)
1 sq. rd = 0.2529 are
1 acre = 0.4047 hectare
1 sq. mile = 2.59 sq. kilometers (km^2)

Table Weights
Following are some common metric measurements of table measure:

1 gram = 0.0527 ounce (oz)
1 kilogram (kg) = 2.2046 lbs
1 metric ton = 1.1023 Eng. ton
1 ounce (oz.) = 28.35 grams
1 lb = 0.4536 kilogram (kg)
1 Eng. ton = 0.9072 metric ton

Approximate Metric Equivalents
Following are some common metric equivalents:

    1 decimeter = 4 inches

    1 meter = 1.1 yards

    1 kilometer = $5/8$ mile

    1 liter = 1.06 qt. liquid; 0.9 qt dry

    1 hectoliter = $2^{5}/8$ bushel

    1 hectare = $2^{1}/2$ acres

    1 stere or cu. meter = $1/4$ cord

    1 kilogram = $2^{1}/5$ lbs.

    1 metric ton = 2200 lbs.

Long Measure
Following are some common measurements of long distance:

    12 inches (in or ″) = 1 foot (ft or ′)

    3 feet = 1 yard (yd) or 36 inches

    $5^{1}/2$ yards or $16^{1}/2$ feet = 1 rod (rd)

    40 rods = 1 furlong (fur) or 660 feet

    8 furlongs or 320 rods = 1 statute mile (mi) or 5280 feet

Nautical Measure
Following are some common equivalents of nautical measure:

    6050.26 ft or 1.15156 statute miles = 1 nautical mile

    3 nautical miles = 1 league

    60 nautical miles or 69.168 statute miles = 1 degree (at the equator)

    360 degrees = circumference of Earth at the equator

Square Measure
Following are some common equivalents of square measure:

    144 square inches (sq. in.) = 1 square foot (sq. ft.)

    9 sq. ft. = 1 square yard (sq. yd.)

    $30^{1}/4$ sq. yd. = 1 square rod (sq. rd.) or perch (P)

    4840 sq. yds. = 1 acre

    640 acres = 1 square mile (sq. mi.)

Mensuration

Mensuration is the process of measuring objects that occupy space (for example, finding the length of a line, area of a triangle, volume of a cube, and so on).

Triangles

Figures bounded by three sides are called *triangles*. Because of variances in the angles and length of sides, there are numerous kinds of triangles.

To find the length of the hypotenuse of a right triangle:
Rule: The hypotenuse is equal to the square root of the sum of the squares of each leg of a triangle.

To find the length of either leg of a right triangle:
Rule: Either leg is equal to the square root of the difference between the square of the hypotenuse and the square of the other leg.

To find the area of any triangle:
Rule: Multiply the base by half the perpendicular height. Thus, if the base is 12 feet and the height 8 feet, the area = $\frac{1}{2}$ of 8 × 12 = 48 sq. ft.

Circle

The Greek letter π (called *pi*) is used to represent 3.1416, the circumference of a circle whose diameter is 1. The circumference of a circle equals the diameter multiplied by 3.1416. The reason for using the decimal 0.7854 to calculate the area of a circle is shown in Figure 3-1.

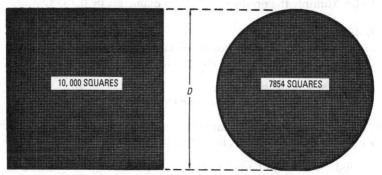

Figure 3-1 The decimal 0.7854 is used to find the area of a circle.

To find the circumference of a circle:
Rule: Multiply the diameter by 3.1416.

To find the diameter of a circle (circumference given):
Rule: Divide the circumference by 3.1416.

NOTE

The numerical equivalent for pi (π) never ends. It can be taken out beyond 1 million places after the decimal point. Most calculators stop at 3.1415927.

To find the area of a circle:
Rule: Multiply the square of the diameter by 0.7854 ($\pi/4$).

To find the diameter of a circle (area given):
Rule: Extract the square root of the area divided by 0.7854.

To find the area of a sector of a circle:
Rule: Multiply the arc of the sector by half the radius.

To find the area of a segment of a circle:
Rule: Find (1) the area of the sector that has the same arc and (2) the area of the triangle formed by the radii and chord. Take the sum of these areas if the segment is greater than 180°; take the difference if less.

To find the area of a ring:
Rule: Take the difference between the areas of the two circles.

To find the area of an ellipse:
Rule: Multiply the product of the two diameters by 0.7854.

Solids

Finding the volume of solids involves the multiplication of three dimensions: length, breadth, and thickness.

To find the volume of a solid:
Rule: Multiply the area of the base by the perpendicular height.

To find the volume of a rectangular solid:
Rule: Multiply the length, breadth, and height.

To find the surface of a cylinder:
Rule: Multiply 3.1416 by the diameter times the length.

To find the volume of a cylinder:
Rule: Multiply 0.7854 by the diameter of the base squared times the length of the cylinder.

To find the surface of a sphere:
Rule: Multiply the area of its great circle by 4.

To find the volume of a sphere:
Rule: Multiply 0.7854 by the cube of the diameter, and then take two-thirds of the product.

To find the volume of a segment of a sphere:
Rule: To three times the square of the radius of the segment's base, add the square of the depth or height. Then, multiply this sum by the depth, and the product by 0.5236.

To find the surface of a cylindrical ring:
Rule: To the thickness of the ring, add the inner diameter. Multiply this sum by the thickness, and the product again by 9.8696.

To find the volume of cylindrical ring:
Rule: To the thickness of the ring, add the inner diameter. Multiply this sum by the square of the thickness, and the product again by 2.4674.

To find the slant area of a cone:
Rule: Multiply 3.1416 by the diameter of the base and by one-half the slant height.

To find the slant area of the frustum of a cone:
Rule: Multiply half the slant height by the sum of the circumferences.

To find the volume of a cone:
Rule: Multiply the area of the base by the perpendicular height, and then multiply by one-third.

To find the volume of a pyramid:
Rule: Multiply the area of the base by one-third of the perpendicular height.

To find the volume of a rectangular solid:
Rule: Multiply the length, breadth, and thickness.

To find the volume of a rectangular wedge:
Rule: Find the area of one of the triangular ends and multiply by the distance between ends.

Mensuration of Surfaces and Volumes

Following are some common formulas for calculating surfaces and volumes:

Area of rectangle = length × breadth

Area of triangle = base × ½ perpendicular height

Diameter of circle = radius × 2

Circumference of circle = diameter × 3.1416

Area of circle = square of diameter × 0.7854 or radius squared × 3.1416

Area of sector of circle:

$$\frac{\text{area of circle} \times \text{number of degrees in arc}}{360}$$

Area of surface of cylinder = circumference × length + area of two ends

To find the diameter of a circle having a given area:
Rule: Divide area by 0.7854, and extract the square root.

To find the volume of a cylinder:
Rule: Multiply the area of the section in square inches by the length in inches to find the volume in cubic inches. Cubic inches divided by 1728 equals the volume in cubic feet.

Surface of a sphere = square of diameter × 3.1416

Volume of a sphere = cube of diameter × 0.5236

Side of an inscribed cube = radius of a sphere × 1.1547

Area of the base of a pyramid or cone, whether round, square, or triangular, multiplied by ⅓ of its height = volume

The following are some typical example questions and their solutions.

Example: What is the radius of a circle the area of which is equal to that of a rectangle whose sides are 26.9 and 12.5 inches, respectively?

Solution:

> Area of rectangle $= 26.9 \times 12.5 = 336.25$ sq. in
> Area of circle $= \pi R^2 = 336.25$ sq. in
>
> From which $R = \sqrt{\dfrac{336.25}{\pi}} = 10.34$ in.

Example: What is the weight of a solid ball of brass 6 inches in diameter? Assume specific gravity $= 8.4$.

Solution: The cubic contents of the ball is obtained by the use of the following formula:

$$V = 0.5236 \times 6^3 = 113.1 \text{ cu. in.}$$

The weight of the ball equals the weight of one cubic inch of water times the specific gravity of brass times the circumference of the ball. Therefore,

$$\text{Weight of ball} = 0.0361 \times 834 \times 113.1 = 34.3 \text{ lbs.}$$

Example: Find the height of a cast-iron cone whose weight is 533.4 kilograms and whose diameter at the base is 5 decimeters. The density of cast iron equals 7.22 kilograms per cubic decimeter.

Solution: The cubic content of the cone is

$$\frac{533.4}{7.22} = 73.9 \text{ cu. decimeters}$$

$$\frac{\text{base area} \times H}{3}$$

and the volume of the cone is

$$H = \frac{3V}{A}$$

$$H = \frac{3 \times 73.9}{5 \times 5 \times 0.7854}$$

$$H = 11.3 \text{ decimeters, or } 11.3 \times 3.937 \text{ in}$$

$$H = 44.5 \text{ in (approx.)}$$

Example: In a certain plumbing installations, three pipes have internal diameters of 2, $2\frac{1}{2}$, and 3 inches, respectively. What is the diameter of a pipe having an area equal to the three pipes?

Solution: The areas of the three pipes are as follows:

$$A_1 + A_2 + A_3 = \frac{\pi \times 2^2}{4} + \frac{\pi \times 2.5^2}{4} + \frac{\pi \times 3^2}{4}$$

$$= \frac{\pi}{4}(2^2 + 2.5^2 + 3^2) = \frac{\pi}{4}(4 + 6.25 + 9) = \frac{\pi}{4} \times 19.25 \text{ sq. in}$$

Since the formula for a circular area is $(\pi \times D^2)/4$ we obtain

$$\frac{\pi}{4} \times D^2 = 0.7854 \times 19.25 = 15.12 \text{ sq. in}$$

Therefore,

$$D = \sqrt{15.12 \div 0.7854}$$

$$D = \sqrt{19.25}, \text{ or } 4.39 \text{ in}$$

Remember that the area of any circular pipe is directly proportional to the square of its diameter in inches, so our calculation will be somewhat simplified. We have

$$2^2 + 2.5^2 + 3^2 = D^2$$

or

$$D = \sqrt{19.25} = 4.39 \text{ in}$$

It follows from the foregoing example that the area of a pipe having the same capacity as a 2-inch, 2½-inch, and 3-inch diameter pipe together must be 4.39 inches in diameter, or 4 25/64 inches

Example: If a 1-inch pipe delivers 10 gallons of water per minute, what size pipe will be required to deliver 20 gallons per minute?

Solution: In this problem, it is necessary to find the diameter of a piece of pipe whose area is twice as large as one that is 1 inch in diameter. Since the area for any circular section $= \pi D^2/4$, it follows that the area for the 1-inch pipe is $(\pi \times 1^2)/4 = 0.7854$ sq. in.

Accordingly, a pipe with twice this area has a diameter of $\pi/2$. Thus, $\pi/2 = \pi D^2/4$, which, after rearrangement of terms, gives $D^2 = 2$, or $D = \sqrt{2} = 1.4142$ inches. That is, the required diameter of a pipe to deliver 20 gpm is 1 13/32 inches (approximately).

Example: If the total fall of a house sewer is 24 inches per 120 feet, what is the slope per foot of this sewer?

Solution: Since the sewer has a total length of 120 feet, the slope per foot is $^{24}/_{120}$, or 0.2 inch.

Example: What is the weight of 1 cubic inch of water if it is assumed that 1 cubic foot weighs 62.5 pounds?

Solution: 1 cubic foot contains 12 inches × 12 inches × 12 inches, or 1,728 cu. in. Therefore, 1 cubic inch weighs 62.5 ÷ 1728, or 0.0361 lb. This figure is given in most handbooks and is frequently used to calculate water pressure in tanks, pipes, and so on.

Example: What is the weight of a column of water 12 inches high and 1 inch in diameter?

Solution: Since 1 cubic inch of water weighs 0.0361 lb, a column of water 12 inches high weighs 12 × 0.0361, or 0.4332 lb.

Example: What is the pressure in pounds per square inch at the base of a 10-foot water cylinder?

Solution: The pressure may be found by remembering that the 10-foot water column weighs 10 × 0.4332, or 4.332 psi.

Example: Calculate in pounds per square inch the minimum water pressure at the city water main that is necessary to fill a house tank located on top of a 6-story building, when the inlet to the water tank is located at an elevation of 110 feet above the city water main.

Solution: Since 1 cubic inch of water weighs 0.0361 lb, the minimum pressure required is 12 × 110 × 0.0361, or 47.7 psi.

Example: What is the number of horsepower required to raise 40,000 pounds 200 feet in 5 minutes? (You may neglect losses.)

Solution: By definition, 1 horsepower is equivalent to doing work at a rate of 33,000 lbs per minute. Thus, in the present problem

$$HP = \frac{\text{lb/ft}}{33,000 \times t}$$

where *HP* is the *horsepower required* and *t* is the *time in minutes*. Substituting values, we obtain

$$HP = \frac{\text{lb/ft}}{33,000 \times t} = \frac{40,000 \times 200}{33,000 \times 5} = 48.5 \, HP$$

Example: What is the number of horsepower required to lift 10,000 gallons of water per hour to a height of 90 feet? (You may neglect losses.) Assume the weight of water to be $8^{1}/_{3}$ lbs per gallon.

Solution:

$$HP = \frac{\text{lb/ft}}{33,000 \times t} = \frac{10,000 \times 8\frac{1}{3} \times 90}{33,000 \times 60} = 3.79 \, HP$$

Example: A city of 25,000 uses 15 gallons of water per day per capita. If it is required to raise this water 150 feet, what is the number of horsepower required? (You may neglect losses.)

Solution:

$$HP = \frac{\text{lb/ft}}{33,000 \times t}$$

$$HP = \frac{15 \times 8\frac{1}{3} \times 25,000 \times 150}{33,000 \times 60 \times 24} = 10 \, HP \, (approx.)$$

Example: How many gallons of water can a 75-horsepower engine raise 150 feet in 5 hours? One gallon of water weighs $8\frac{1}{3}$ lbs. (You may neglect losses.)

Solution: If the given data are substituted in our formula for horsepower we obtain:

$$75 = \frac{150 \times 8\frac{1}{3} \times G}{33,000 \times 5 \times 60}$$

$$G = \frac{75 \times 33,000 \times 5 \times 60}{150 \times 8\frac{1}{3}} = 594,000 \text{ gallons}$$

Example: A circular tank 20 feet deep and 20 feet in diameter is filled with water. If the average height to which the tank is to be lifted is 50 feet, what must be the horsepower of an engine capable of pumping the water out in 2 hours? (You may neglect losses.)

Solution: In this example, it is first necessary to calculate the cubic content of the tank (that is, the cross-sectional area multiplied by its height). Thus,

Volume in cubic feet $= \pi R^2 H = \pi \times 10^2 \times 20 = 6283$ cu. ft

Since 1 cubic foot of water weighs 62.5 lbs, the total weight of the tank's contents is

$62.5 \times 6283 = 392,700$ pounds (approx.)

Again, using our formula, we have

$$HP = \frac{\text{lb/ft}}{33,000 \times t} = \frac{392,700 \times 50}{33,000 \times 2 \times 60} = 5 \, HP \, (approx.)$$

Example: The suction lift on a pump is 10 feet and the head pumped against is 100 feet. If the loss caused by friction in the pipe line is assumed as 9 feet, and the pump delivers 100 gpm, what is the horsepower delivered by the pump?

Solution: When the water delivered is expressed in gpm the formula for horsepower is

$$HP = \frac{GPM \times \text{head in feet} \times 8.33}{33,000}$$

Then,

$$HP = \frac{GPM \times 119 \times 8.33}{33,000}$$

A substitution of values gives

$$HP = \frac{100 \times 119 \times 8.33}{33,000} = 3\ HP$$

Example: A tank having a capacity of 10,000 gallons must be emptied in 2 hours. What capacity pump is required?

Solution: The capacity of the pump in gpm is arrived at by dividing the total gallonage of the tank by the time in minutes. Thus,

$$GPM = \frac{10,000}{2 \times 60} = 83.3\ GPM$$

Example: What is the capacity of a double-acting pump in gpm when it makes 60 strokes per minute if the cylinder is 9 inches in diameter and the stroke is 10 inches? (You may neglect slippage.)

Solution: The capacity of the pump in gpm is

$$GPM = \frac{D^2 \times 0.7854 \times L \times N}{231}$$

$$= \frac{9^2 \times 0.7854 \times 10 \times 60}{231}$$

$$= 165.2\ GPM$$

Example: What is the indicated horsepower of a 4-cylinder 5 × 6 engine running at 500 rpm and 50 psi effective pressure?

Solution: The well-known formula for calculation of indicated horsepower is

$$IHP = \frac{PLAN}{33,000} \times K$$

where
P = Effective pressure in psi acting on the piston
L = Length of stroke in feet
A = Area of piston in sq. in
N = Number of strokes per minute
K = Coefficient equal to $\frac{1}{2}$ times number of cylinders

Substituting our values, we obtain

$$IHP = \frac{50 \times \frac{6}{12} \times 0.7854 \times 5^2 \times 500}{33,000} \times 2 = 14.87$$

Example: The temperature of a boiler as registered by a pyrometer is 2750°F. What is the corresponding reading in Centigrade degrees?

Solution: The equation is

$$C = \frac{5}{9}(F - 32)$$
$$C = \frac{5}{9}(2750 - 32) = 1510°C$$

Example: A pail contains 58 lbs of water having a temperature of 40°F. If heat is applied until the temperature of the water reaches 95°F, what is the amount of Btu supplied to the water?

Solution: The rise in temperature has been 95–40 = 55°F. Since one Btu is one pound raised one degree, it follows that to raise 58 pounds 55 degrees requires

$$58 \times 55 = 3190 \text{ Btu}$$

Example: How many heat units (Btu) are required to raise one pound of water from 55° to 212°F? How many ft-lbs of work does this represent?

Solution: The number of heat units required is

$$1(212 - 55) = 157 \text{ Btu}$$

Since the mechanical equivalent in ft-lbs of heat is 778, it is only necessary to multiply the number of heat units by this constant to obtain the equivalent number of work units. Thus,

$$157 \times 778 = 122,146 \text{ ft-lbs}$$

Example: In a certain pump installation, it was found that, because of corrosion, a 2-inch pipe had an effective diameter of only 1½ inches. Calculate the loss in cross-sectional area caused by corrosion.

Solution: It may be easily shown that the area of any circular pipe varies as the square of its diameter. The loss in cross-sectional area is therefore

$$(2 \times 2) - (1.5 \times 1.5) = 4 - 2.25, \text{ or } 1.75$$

Since the cross-sectional area formula is $0.7854 \times D^2$, then loss in the cross-sectional area $= 0.7854 \times 1.75 = 1.37$ sq. in.

Methods for Calculating Pipe Offset Changes (Other Than 90°)

Figure 3-2 illustrates a pipe change of direction where it is necessary to change the position of pipeline L to a parallel position F to avoid some obstruction such as wall E. When the two lines L and F are to be fitted with elbows having an angle of 90°, the plumber must find the length of the pipe H connecting the two elbows A and C. The distance BC must also be determined in order to fix the point A so that elbows A and C will be in alignment. There are several methods of solving this problem, of which two follow.

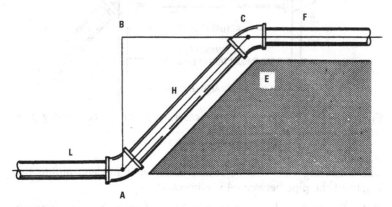

Figure 3-2 A pipeline connection of 45-degree elbows and the method of finding lengths of connecting pipe.

Method I

In the triangle ABC,

$$(C)^2 = (AB)^2 + (BC)^2$$

from which

$$AC = \sqrt{(AB)^2 + (BC)^2}$$

Example: If the distance between pipelines L and F in Figure 3-2 is 20 inches (offset AB), what length of pipe (H) is required to connect elbows A and C? When 45° elbows are used, both offsets are equal. Thus, substituting in the equation

$$AC = \sqrt{20^2 + 20^2} = \sqrt{800} = 28.28 \text{ in.}$$

The length of the pipe just calculated does not allow for the projections of the elbows.

This must be taken into account, as shown in Figure 3-3.

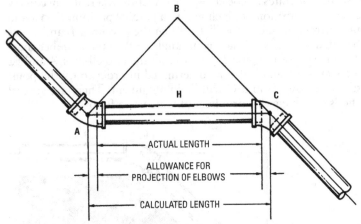

Figure 3-3 The calculated and actual length of the connecting pipe with elbow other than 90 degrees.

Method 2
The following rule will be found convenient in determining the length of the pipe between 45° elbows.

Rule: For each inch of offset, add $^{53}/_{128}$ of an inch, and the result will be the length between centers of the elbows.

Example: Calculate the length AC (Figure 3-2) by the preceding rule:

$$20 \times {}^{53}/_{128} = {}^{1060}/_{128} = 8^9/_{32}$$

Add this to the offset:

$$20 + 8\,{}^9\!/_{32} = 28\,{}^9\!/_{32}$$

This is the calculated length; to obtain the actual length, deduct the allowance for the projection of the elbows, as in Figure 3-3.

Calculators

A calculator can be used to improve the accuracy and the speed of calculations. They are inexpensive and easy to operate. The scientific type is best for plumbing work since it has the square root and the reciprocal as well as *pi*, square, and raising to a power.

All trigonometric functions are also available for easy use with the Radio Shack or Hewlett-Packard or Texas Instruments models. Most of the HP and TI models have sufficient functions to solve most plumbing problems.

Differences in Calculators

The Hewlett-Packard varies slightly from the Texas Instrument calculator. The Hewlett-Packard has reverse Polish notation. This means you enter the first number and then follow the formula as written. The Texas Instrument procedure is slightly different. It is suggested that you follow the instruction booklet closely before attempting to perform the calculations.

The calculator you use should have scientific tables, with the ability to do the following:

Function	Key
Square Root	$[\sqrt{X}]$
Reciprocals	$[1/x]$
Squares	$[x^2]$
Trig Functions (cosine)	$[\cos]$
Add	$[+]$
Subtract	$[-]$
Divide	$[\div]$
Multiply	$[\times]$

The calculator also should have at least one level of memory.

It is helpful if the calculator computes *pi* (π). The solutions to problems can then be more accurate if *pi* is taken out to six or nine places. On most calculators *pi* is equal to 3.14159217 and is rounded to 3.1416 in most problems. The number representing *pi* never comes out even. It has been taken out to more than 1 million places past the decimal point and still it does not come out even.

Conversions

In some instances, plumbing measurements are given in fractions of an inch. To convert a fraction to a decimal so it can be used with a calculator, just divide the numerator (top number) by the denominator (bottom number). For example, $1/3$ is 0.333, or divide 1 by 3. The answer is 0.33333333 on the calculator. In like fashion, $3/8$ is 3 divided by 8, or 0.375. See Table 3-6 for some typical fractions and their decimal equivalents.

Table 3-6 Decimal Equivalents

Fraction	Decimal	Fraction	Decimal
$1/32$	0.03125	$9/16$	0.56250
$1/16$	0.06250	$19/32$	0.59375
$3/32$	0.09375	$5/8$	0.62500
$1/8$	0.12500	$21/32$	0.65625
$5/32$	0.15625	$11/16$	0.68750
$3/16$	0.18750	$23/32$	0.71875
$1/4$	0.25000	$3/4$	0.75000
$9/32$	0.28125	$25/32$	0.78125
$5/16$	0.31250	$13/16$	0.81250
$11/32$	0.34375	$27/32$	0.84375
$3/8$	0.37500	$7/8$	0.87500
$13/32$	0.40625	$29/32$	0.90625
$7/16$	0.43750	$15/18$	0.93750
$15/32$	0.46875	$31/32$	0.96875
$1/2$	0.50000	1	1.000000
$17/32$	0.53125		

On the calculator, you can just enter the bottom number and press the "1/x" key (the reciprocal key) to produce an instant answer (the reciprocal).

Multiple Choice Exercises

Select the right answer and blacken $a, b, c,$ or d on your practice answer sheet. (See Appendix C for answers.)

1. Which of these symbols is used to indicate square root?

    a. X

    b. ÷

    c. $\sqrt{}$

    d. ±

2. In addition to common math symbols, the plumber should be familiar with the diameter, degrees, square feet, and square inch. Which of the following symbols represents square feet?

    a. □′

    b. ≤

    c. Ø

    d. ∇

3. Cubic measure is used to represent _____.

    a. volume

    b. area

    c. quantity

    d. power

4. In order to change a fraction to a decimal you must _____.

    a. divide the numerator by the denominator

    b. divide the denominator by the numerator

    c. multiply the denominator by the numerator

    d. add the denominator and the numerator

5. Using the squares, cubes, square root, and cube root tables, find the square root of 50.

    a. 3.68403

    b. 7.07106

    c. 8.00110

    d. 7.14142

6. A millimeter is the same as _____.

    a. 1 inch

    b. 0.03937 inch

    c. 0.1 of a meter

    d. 1 myriameter

7. Which Greek letter is used to represent 3.1416?

   a. ϕ

   b. φ

   c. μ

   d. π

8. Finding the volume of a solid involves the _____ of three dimensions (namely, length, breadth, and thickness).

   a. division

   b. multiplication

   c. square root

   d. addition

9. Using the Table 3-6, convert $5/32$ to a decimal.

   a. 0.3125

   b. 0.68750

   c. 0.15625

   d. 0.25000

10. The circumference of a circle is found by multiplying the diameter by _____.

   a. 0.25000

   b. Greek letter pi

   c. Greek letter omega

   d. Greek letter alpha

Chapter 4

Materials and Fittings

In any trade or profession, information is critical. Information is needed to make correct decisions when solving problems and designing systems. In this chapter, you will learn how to read abbreviations and/or acronyms for various organizations that you will encounter as a plumber.

Code requirements are of utmost importance to knowing the trade. Tables contain information on various materials and fittings. They present that information in a concise manner. Learning to read a table is a necessary part of the trade. Testing your ability to understand what you learn from a table is part of the license examination.

The following are typical questions asked in reference to materials and fittings in plumbing licensing examinations.

4-1 What are the usual plumbing code requirements for material installed in plumbing systems?

The plumbing codes from which many journeymen and master plumbing examination questions are taken require that materials used in drainage or plumbing systems (or any part of such systems) be free of defects that would impair service or result in unsanitary conditions. In addition, it is generally required that each length of pipe, trap, fitting, and fixture used in the plumbing or drainage system be cast, stamped, or indelibly marked with the maker's name or registered trademark. Moreover, methods, materials, and fixtures must meet established technical standards of quality so as to produce safe and sanitary plumbing installations and conditions.

General available standards or specifications are usually referred to in order to standardize and ensure safe and acceptable fixtures and materials. Table 4-1 lists various institutions with sources of reliable technical information and requirements as referred to by various state and local code administrative authorities. Code administrative personnel normally update established technical standards at two-year intervals.

4-2 May any thickness of sheet lead or fittings be used for installation in plumbing systems?

Sheet lead shall not weigh less than 4 pounds per square foot, and lead bends and lead traps shall not be less than $1/8$ inch in wall thickness.

Table 4-1 Standards and Specification Sources with Organization Abbreviations

Abbreviation	Organization
AGA	American Gas Association, Inc.
ANSI	American National Standards Institute, Inc.
ASME	American Society of Mechanical Engineers
ASSE	American Society of Sanitary Engineers
ASTM	American Society for Testing and Material
AWWA	American Water Works Association
CS	Commercial Standards, Supt. of Documents
FS	Federal Specifications, U.S. General Services Administration
MSS	Manufacturers Standardization Society of Valve and Fittings Industry
NSF	National Sanitation Foundation Testing Laboratories, Inc.
SPR	Simplified Practice Recommendations representing recorded recommendations of the trade and issued by the U.S. Department of Commerce
USAI	USA Standards approved by the USA Standards Institute
UL	Underwriters' Laboratories, Inc.
WCF	Water Conditioning Foundation

4-3 Are there special copper sheet thickness requirements when used in conjunction with plumbing safe pans or vent flashings?

Sheet copper used for safe pans is recommended to be not less than 12 oz per square foot, and for vent terminal flashings not less than 8 oz per square foot. In all instances, material shall be of sufficient weight to serve the purpose for which it is used.

4-4 Does gage thickness vary for different sizes of galvanized metal plumbing ductwork?

Sheet iron when used for pipe ductwork with plumbing installations is recommended to have a minimum gage thickness as follows:

Number 26 gage for 2- to 12-inch pipe

Number 24 gage for 13- to 20-inch pipe

Number 22 gage for 21- to 26-inch pipe

4-5 Are there recommended weights and sizes for brass caulking ferrules for plumbing work installation?

Brass caulking ferrules shall be of red brass or heavy cast red brass with weights and dimensions in accordance with Table 4-2.

Table 4-2 General Plumbing Code Recommended Weights and Sizes for Brass Caulking Ferrules

Pipe Size	Actual Inside Diameter	Length	Weight
2 inches	$2^{1}/_{4}$ inches	$4^{1}/_{2}$ inches	1 lb, 0 oz
3 inches	$3^{1}/_{4}$ inches	$4^{1}/_{2}$ inches	1 lb, 12 oz
4 inches	$4^{1}/_{4}$ inches	$4^{1}/_{2}$ inches ·	2 lbs, 8 oz

4-6 What are the recommended weights and sizes for soldering bushings in plumbing systems?

Soldering bushings for plumbing installations are recommended to be of red brass pipe or of heavy cast red brass of a standard size or of dimensions listed in Table 4-3.

Table 4-3 General Recommended Weights and Sizes for Plumbing Brass Soldering Bushings

Pipe Size (inches)	Minimum Weight (each per foot)	Pipe Size (inches)	Minimum Weight (each per foot)
$1^{1}/_{4}$	6 oz	$2^{1}/_{2}$	1 lb, 6 oz
$1^{1}/_{2}$	8 oz	3	2 lbs
2	14 oz	4	3 lbs, 8 oz

4-7 Does a minimum thickness requirement exist for water-flushed toilet floor and wall flanges?

Recommended floor- and wall-flange thickness for water-flushed closets are that they not be less than $^{3}/_{16}$ inch thick when cast iron and not less than $^{1}/_{8}$ inch thick when of brass.

4-8 Must cleanout plugs for drain piping be of brass?

Cleanout plugs shall be of brass conforming to Federal Specifications WW-P-401. Plugs may have either raised square or countersunk heads. Countersunk heads are recommended for use where raised heads may result in a hazard. Figure 4-1 illustrates a typical cleanout plug in cast-iron bell-type sewer pipe. Some codes allow plastic pipe and plastic cleanout plugs.

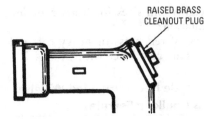

RAISED BRASS
CLEANOUT PLUG

Figure 4-1 A typical cleanout used in cast-iron bell-type sewer pipe.

4-9 Are there nationally established testing or technical standards for the various types of pipes used in plumbing work?

There are nationally recognized standards for plumbing piping and fittings as approved by the various testing laboratories and technical organizations. Following are some of the approved piping specification standards.

- *Cast-iron pipe*—Cast-iron pipe used for waste plumbing should conform to ASTM "Standard Specifications for Cast-Iron Soil Pipe and Fittings" and be extra heavy.
- *Wrought-iron pipe*—All wrought-iron pipe shall be galvanized conforming to ASTM "Standard Specifications for Welded Wrought-Iron Pipe" (serial designation A 72-33).
- *Mild-steel pipe*—Mild-steel pipe shall be galvanized conforming to ASTM "Standard Specifications for Welded and Seamless Steel Pipe" (serial designation A 53-33).
- *Copper and brass pipe*—Many plumbing code administrative authorities recommend that copper pipe conform to ASTM "Standard Specifications" (B 42-33), copper water tubing to ASTM "Standard Specifications" (B 88-33), and brass pipe to ASTM "Standard Specifications" (B 43-33). In addition, most examination questions will recommend that thicknesses for brass or copper concealed waste pipes and traps shall not be less than 14 B&S gage thickness.

4-10 By definition, what is a pipe nipple?

A *pipe nipple* is a piece of pipe less than 12 inches in length threaded on both ends. Pipe longer than 12 inches is regarded as *cut pipe*. With respect to length, nipples (as shown in Figure 4-2) may be classified as follows:

- Close
- Short
- Long

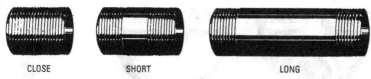

CLOSE   SHORT   LONG

Figure 4-2 Typical pipe nipples.

Where fittings or valves are to be very close to each other, the intervening nipple is just long enough to take the threads at each end; this is called a *close nipple.* If a small amount of pipe exists between the threads, it is called a *shoulder* or *short nipple,* and where a larger amount of bare pipe exists, it is called a *long nipple* or *extra-long nipple.*

4-11 Describe a pipe coupling.

Ordinary couplings often come with the pipe, one coupling to each length (see Figure 4-3). The couplings are made of wrought or cast metal, or of brass. They are regularly threaded with right-hand threads but can be obtained on special order with right- and left-hand threads. Another form of coupling, called an extension piece, differs from the standard coupling in that it has a male thread at one end (see Figure 4-4).

4-12 What are back- and side-outlet return bends?

These are return bends that have been provided with an additional outlet at the back or side (see Figure 4-5). They are regularly made in sizes ranging from ³/₄ inch to 3 inches, inclusive, in the closed or open patterns.

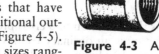

Figure 4-3 A pipe coupling.

4-13 Which are the most important of the branching pipe fittings?

Tees are the most important and widely used branch fittings. Tees, like elbows, are made in a multiplicity of sizes and patterns. They are used for making a branch of 90° to the main pipe, and always have the branch at

Figure 4-4 A pipe extension.

right angles. When the three outlets are of the same size, the fitting is specified by the size of pipe (such as ¹/₂-inch tee). When the branch is a different size than the run outlets, the size of the run is given first (such as 1-inch × ¹/₄-inch tee). When all three outlets are of different

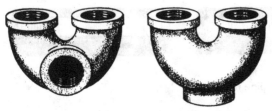

Figure 4-5   Cast-iron return bends with back and side outlets.

sizes, they are all specified, giving the sizes of run first (such as a 1¼-inch × 1-inch × 1½-inch tee).

The method of specifying tees is illustrated in Figure 4-6. To avoid mistakes in ordering tees that do not have all outlets of the same size, make diagrams with dimensions as shown in the illustration. Figure 4-7 shows various tees and illustrates the great variety of patterns in use.

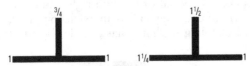

Figure 4-6   A method of specifying dimension of tees.

4-14 What is the definition of a cross plumbing fitting?

A *cross fitting* is an ordinary tee having a back outlet opposite the branch outlet (see Figure 4-8). The axes of the four outlets are in the same plane and at right angles to each other. Crosses, like tees, are made in a number of sizes.

4-15 What is the purpose of a plug when used in plumbing work?

Plugs are used for closing the end of a pipe or a fitting that has a female thread. Plugs are made of cast iron, malleable iron, and brass. Figure 4-9 shows the various patterns. Usually a square head or a four-sided countersunk head is used for the small sizes; a hexagon head is used for the larger sizes. Normal plug sizes range from ⅛ inch to 12 inches, inclusive.

4-16 Do union types vary?

Types of unions vary. A *plain union,* as shown in Figure 4-10, requires a gasket. The two pipes to be joined by the union must be

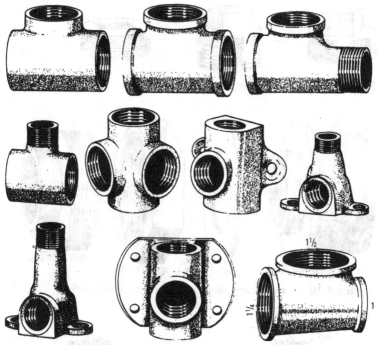

Figure 4-7   Various cast-iron tees.

Figure 4-8   Two typical cross connections.

in approximate alignment to secure a tight joint because of the flat surfaces that must press against the gasket. This limitation is shown in Figure 4-11.

To avoid alignment difficulties, as well as to avoid problems with the gasket, unions with spherical seats and ground joints are now available. These consist of a composition ring bearing against iron, or with both contact surfaces of composition. Figure 4-12 shows the

Figure 4-9 Various pipe plugs.

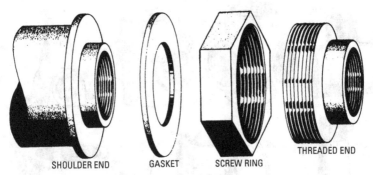

SHOULDER END   GASKET   SCREW RING   THREADED END

Figure 4-10 An ordinary joint union that requires a gasket.

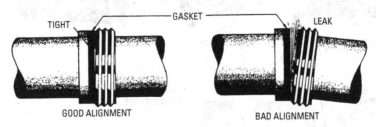

TIGHT   GASKET   LEAK

GOOD ALIGNMENT   BAD ALIGNMENT

Figure 4-11 An illustration of the importance of proper alignment.

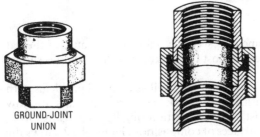

GROUND-JOINT UNION

Figure 4-12 An ordinary ground-joint union.

construction of a ground-joint union. The joint has spherical contact, and the illustration shows the tight joint secured, even though the pipes are out of line. There are also unions made entirely of brass with ground joints.

4-17 What are the functions of bushings in plumbing systems and how are they listed for pipe sizing?

The function of a bushing is to connect the male end of a pipe to a fitting of a larger size. It consists of a hollow plug with male and female threads to suit the different diameters. A bushing may be regarded as either a *reducing* or an *enlarging fitting*. Bushings $2\frac{1}{2}$ inches and smaller, reducing one size, are usually malleable iron; reducing two or more sizes are cast iron, except brass bushings, which may be obtained in sizes from $\frac{1}{4}$ inch to 4 inches.

Bushings are listed by the pipe size of the male thread; thus, a $\frac{1}{4}$-inch bushing joins a $\frac{1}{4}$-inch fitting to a $\frac{1}{8}$-inch pipe. To avoid mistakes, however, it is better to specify the size of both threads (for example, by calling the bushing just mentioned a $\frac{1}{4}$-inch $\times$ $\frac{1}{8}$-inch bushing). The regular pattern bushing has a hexagon nut at the female end for screwing the bushing into the fitting. The faced bushing is used for very close work and has a faced end in place of the hexagon nut. This may be used with a long screw pipe and faced locknut to form a tight joint or to receive a male end fitting for close work. Figure 4-13 shows the plain and faced type of bushings.

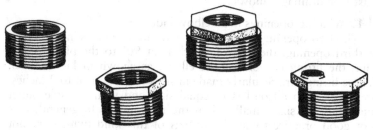

Figure 4-13   Various bushings.

4-18 List some of the uses of elbows in plumbing.

Elbows are used where it is necessary to change the direction of a pipeline in any of several standard and special angles. For gas, water, and steam, the standard angles are 45° and 90°, while the special angles are $22\frac{1}{2}$° and 60°. Cast-iron drainage fitting elbows are regularly made with angles of $5\frac{5}{8}$°, $11\frac{1}{4}$°, $22\frac{1}{2}$°, 45°, 60°, and

90°. Elbow angles measure the degree that the direction is changed, as shown in Figure 4-14. These figures should be carefully noted. The angle is not the angle between the two arms, but rather the angle between the axis of one arm and the projected axis of the other arm, as shown in the illustration.

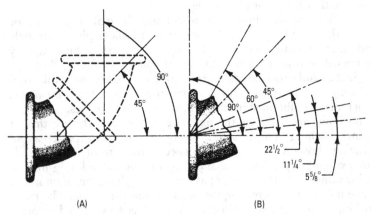

(A)   (B)

Figure 4-14   Standard elbow angles.

There are a large variety of elbows. Figure 4-15 shows some standard patterns, which include reducing cast-iron elbows and various cast-iron drainage elbows.

4-19 What do openings in an elbow indicate?

The two openings of an elbow indicate its run. When there is a third opening, the axis of which is at 90° to the plane of the run, the fitting is a side-outlet elbow, as shown in Figure 4-16. These fittings are regularly made in sizes from 1/4 inch to 2 inches, inclusive, with all outlets of equal size, and with the side outlet one and two sizes smaller than the rim outlets. In general, it is not good practice to specify fittings of this kind (which are not too much in demand), unless the more usual forms are difficult to find.

4-20 How are pipe caps used in a plumbing system?

Pipe caps are used for closing the end of a pipe or fitting that has a male thread. Caps, like plugs, are made of cast iron, malleable iron, and brass. Figure 4-17 shows various cap designs. Plain and flat-band or beaded caps are regularly made in sizes from 1/8 inch to 6 inches, inclusive; cast-iron caps from 3/8 inch to 15 inches, inclusive.

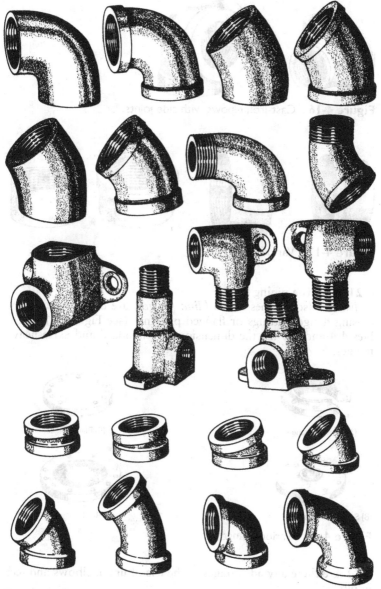

Figure 4-15 Various cast-iron elbows.

Figure 4-16 Cast-iron elbows with side joints.

Figure 4-17 Various pipe caps.

4-21 What is a piping flange?

Flanges (sometimes called *blind flanges*) are cast-iron discs for closing flanged fittings or flanged pipelines (see Figure 4-18). Tables 4-4 and 4-5 list the dimensions for standard and extra-heavy flanges.

(A) Screwed flange.

(C) Blind flange.

(B) Screwed reducing flange.

(D) Blind flange.

Figure 4-18 Various flanges.

4-22 Are there any advantages in the use of union elbows and tees in pipelines?

The frequent use of unions in pipelines is desirable for convenience in case of repairs. Where the union is combined with a fit-

Table 4-4 Sizes and Dimensions of Screwed and Blind Flanges (see Figure 4-18)

Pipe Size and Diameter (inches)	Thickness (inches)			Bolt Sizes	
	Through Flange	Through Hub	Through Flange	Number of Holes	Dia. of Bolts (inches)
	Flange A		Flange C & D	Holes	
1 × 4¼	$7/16$	$11/16$	$7/16$	4	$1/2$
1¼ × 4⅝	$1/2$	$13/16$	$1/2$	4	$1/2$
1½ × 5	$9/16$	$7/8$	$9/16$	4	$1/2$
2 × 6	$5/8$	1	$5/8$	4	$5/8$
2½ × 7	$11/16$	$13/16$	$11/16$	4	$5/8$
3 × 7½	$3/4$	$1 1/4$	$3/4$	4	$5/8$
3½ × 8½	$13/16$	$1 1/4$	$13/16$	8	$5/8$
4 × 9	$15/16$	$1 15/16$	$15/16$	8	$5/8$
5 × 10	$15/16$	$1 7/16$	$15/16$	8	$3/4$
6 × 11	1	$1 9/16$	1	8	$3/4$
8 × 13½	$1 1/8$	$1 3/4$	$1 1/8$	8	$3/4$
10 × 16	$1 3/16$	$1 15/16$	$1 3/16$	12	$7/8$
12 × 19	$1 1/4$	$2 3/16$	$1 1/4$	12	$7/8$

ting, the advantage of a union is obtained with only one threaded joint instead of two, as in the case of a separate union. Figure 4-19 shows various union elbows and union tees of the male and female types.

4-23 Must consideration be given to linear expansion and contraction of pipe caused by differences in the temperature of the fluid carried and the surrounding air?

The linear expansion and contraction of pipe, due to differences in temperature of the fluid carried and the surrounding air, must be cared for by suitable expansion joints or bends. Table 4-6 is provided in order to determine the amount of expansion or contraction in a pipeline, showing the increase in length of a pipe 100 feet long at various temperatures. The expansion of any length of pipe may be found by taking the difference in increased length at the minimum and maximum temperatures, dividing by 100, and multiplying by the length of the line under consideration.

Table 4-5 Sizes and Dimensions of Screwed Reducing Flanges

Pipe Size and Dia. (inches)	Thickness Through Flange (inches)	Bolt Sizes		Pipe Size and Dia. (inches)	Thickness Through Flange (inches)	Bolt Sizes	
		Number of Holes	Dia. of Bolts			Number of Holes	Dia. of Bolts
1 × 5	$9/16$	4	$1/2$	3 × 10	$15/16$	8	$3/4$
1 × 6	$5/8$	4	$5/8$	4 × 10	$15/16$	8	$3/4$
$1^1/4$ × 6	$5/8$	4	$5/8$	2 × 11	1	8	$3/4$
$1^1/2$ × 6	$5/8$	4	$5/8$	$2^1/2$ × 11	1	8	$3/4$
$1^1/2$ × 7	$11/16$	4	$5/8$	3 × 11	1	8	$3/4$
2 × 7	$11/16$	4	$5/8$	4 × 11	1	8	$3/4$
$1^1/2$ × $7^1/2$	$3/4$	4	$5/8$	5 × 11	1	8	$3/4$
2 × $7^1/2$	$3/4$	4	$5/8$	3 × $13^1/2$	$1^1/8$	8	$3/4$
$2^1/2$ × $7^1/2$	$3/4$	4	$5/8$	4 × $13^1/2$	$1^1/8$	8	$3/4$
$3^1/2$ × $8^1/2$	$13/16$	8	$5/8$	5 × $13^1/2$	$1^1/8$	8	$3/4$
$1^1/2$ × 9	$15/16$	8	$5/8$	6 × $13^1/2$	$1^1/8$	8	$3/4$
2 × 9	$15/16$	8	$5/8$	6 × 16	$1^3/16$	12	$7/8$
$2^1/2$ × 9	$15/16$	8	$5/8$	8 × 16	$1^3/16$	12	$7/8$
3 × 9	$15/16$	8	$5/8$	8 × 19	$1^1/4$	12	$7/8$
$3^1/2$ × 9	$15/16$	8	$5/8$	10 × 19	$1^1/4$	12	$7/8$

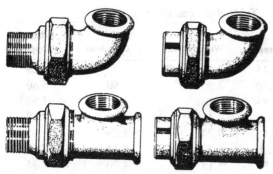

Figure 4-19 Various union elbows and union tees.

Table 4-6 Expansion of Steam Pipes (inches increase per 100 feet)

Temperature (°F)	Steel	Wrought iron	Cast iron	Brass and Copper
0	0	0	0	0
20	0.15	0.15	0.10	0.25
40	0.30	0.30	0.25	0.45
60	0.45	0.45	0.40	0.65
80	0.60	0.60	0.55	0.90
100	0.75	0.80	0.70	1.15
120	0.90	0.95	0.85	1.40
140	1.10	1.15	1.00	1.65
160	1.25	1.35	1.15	1.90
180	1.45	1.50	1.30	2.15
200	1.60	1.65	1.50	2.40
220	1.80	1.85	1.65	2.65
240	2.00	2.05	1.80	2.90
260	2.15	2.20	1.95	3.15
280	2.35	2.40	2.15	3.45
300	2.50	2.60	2.35	3.75
320	2.70	2.80	2.50	4.05
340	2.90	3.05	2.70	4.35
360	3.05	3.25	2.90	4.65
380	3.25	3.45	3.10	4.95

(continued)

Table 4-6   (continued)

Temperature (°F)	Steel	Wrought iron	Cast iron	Brass and Copper
400	3.45	3.65	3.30	5.25
420	3.70	3.90	3.50	5.60
440	3.95	4.20	3.75	5.95
460	4.20	4.45	4.00	6.30
480	4.45	4.70	4.25	6.65
500	4.70	4.90	4.45	7.05
520	4.95	5.15	4.70	7.45
540	5.20	5.40	4.95	7.85
560	5.45	5.70	5.20	8.25
580	5.70	6.00	5.45	8.65
600	6.00	6.25	5.70	9.05
620	6.30	6.55	5.95	9.50
640	6.55	6.85	6.25	9.95
660	6.90	7.20	6.55	10.40
680	7.20	7.50	6.85	10.95
700	7.50	7.85	7.15	11.40
720	7.80	8.20	7.45	11.90
740	8.20	8.55	7.80	12.40
760	8.55	8.90	8.15	12.95
780	8.95	9.30	8.50	13.50
800	9.30	9.75	8.90	14.10

4-24 Why are special fittings needed to connect iron and copper pipes?

To prevent galvanic current from being generated by the two dissimilar metals.

4-25 What is the name of the process in which steel and wrought iron pipe rust?

The name of the process is *tuberculation,* so called because of the bumps or small nodules that are found on the rusting pipe.

Matching Exercises

Match the following items with the lettered artwork. (See Appendix C for answers.)

_____ **1.** Brass cleanout plug

_____ **2.** Typical pipe nipples

_____ **3.** A pipe coupling
_____ **4.** A pipe extension
_____ **5.** Cast-iron return bends
_____ **6.** Cast-iron tees
_____ **7.** Pipe plugs
_____ **8.** Ground-joint union
_____ **9.** Bushings
_____ **10.** Cast-iron elbows
_____ **11.** Cast-iron elbows with side joints
_____ **12.** Flanges
_____ **13.** Pipe caps
_____ **14.** Elbows and union tees
_____ **15.** Cross connections

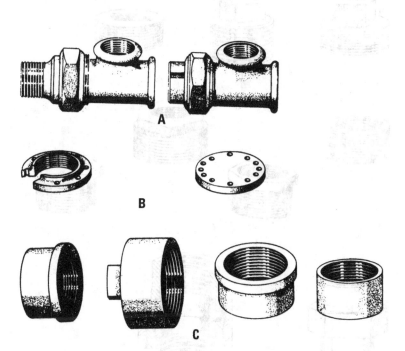

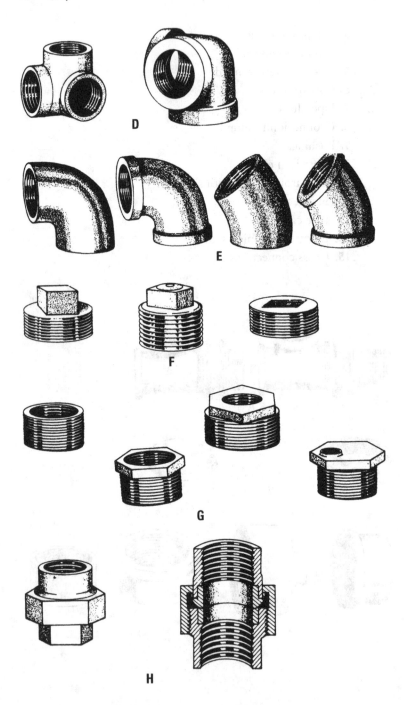

D

E

F

G

H

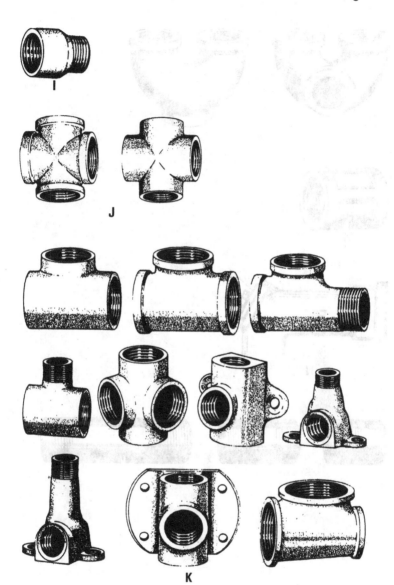

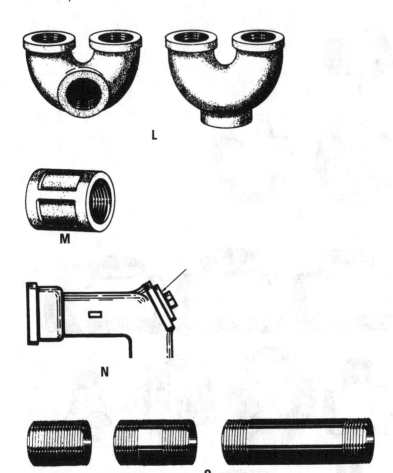

L

M

N

O

Chapter 5

Joints and Connections

All plumbing codes require that joints and connections in the plumbing system be watertight and gas-tight. There are exceptions to water- and gas-tight joints and connections, namely in those sections of the system that are open-joint, or perforated piping that has been installed for the purpose of collection and disposal of ground or seepage water to below-surface storm drainage systems.

The following are typical questions asked in reference to joints and connections in plumbing licensing examinations.

5-1 How are joints of cast-iron bell-and-spigot soil pipe made?

All caulked joints of cast iron bell-and-spigot soil pipe shall be firmly packed with oakum or hemp, and shall be secured only with pure lead to a minimum depth of 1 inch and shall be run in one pouring and caulked tight. Figure 5-1 illustrates a bell-and-spigot joint used for connecting lengths of cast-iron soil pipe. Figure 5-2 illustrates a method used for packing oakum into the bell-and-spigot connection of a sewer pipe.

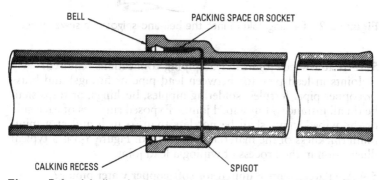

Figure 5-1 A bell-and-spigot joint used for connecting lengths of cast-iron pipe.

The use of jute, hemp, or oakum is not permitted for caulking or for joining water distribution piping connections.

5-2 What are the general recommendations for assembling threaded and screwed pipe joints?

All burrs or cuttings of threaded or screwed joints shall be removed before assembling or joining connections. Pipe joint cement or mixture shall be used only on male threads.

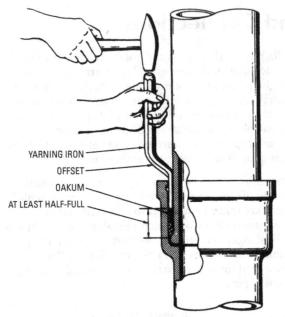

YARNING IRON
OFFSET
OAKUM
AT LEAST HALF-FULL

Figure 5-2 Packing oakum into the bell-and-spigot of a sewer pipe.

5-3 Are full-wiped joints in lead piping connections required?

Joints in lead pipe (or between lead pipe or fittings) and brass or copper pipe, ferrules, soldering nipples, bushings, or traps shall be (in all instances) full-wiped joints. Exposed surfaces of each side of the wiped joints shall be not less than $3/4$ inch and at a minimum thickness of the material being joined. Figure 5-3 is a typical illustration in the process of wiping a lead joint.

5-4 Are flared joints required for soft-copper water piping?

Flared joints are used when making connections in soft-copper water piping. Expansion of the tubing is made with an approved flaring tool in accordance with the best standards and approved practices. Copper tube and fittings are generally required to conform to ASTM "Standard Specifications for Seamless Copper Water Tube" (serial number 3-88 of the latest revision), or local administrative approval authority.

5-5 Are collars required for connecting precast pipe?

Precast pipe joints are connected with collars. These collars are formed in both bell-and-spigot ends of the piping prior to use. Collar

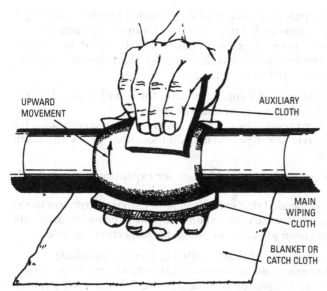

UPWARD
MOVEMENT

AUXILIARY
CLOTH

MAIN
WIPING
CLOTH

BLANKET OR
CATCH CLOTH

Figure 5-3 Wiping a horizontal lead joint.

surfaces shall be slightly conical and, before joint-finishing contact, shall be cleaned and coated with approved adhesives. After adhesive application, the spigot end of the pipe is inserted into the bell, and pressure is applied to ensure proper seating.

5-6 What materials are used for connecting clay pipe joints?

Resilient-type materials are used for connecting lengths of clay pipe in building sewers on both spigot and bell ends. Most codes and referencing code examination questions state that these materials shall meet the standards established by Type III of ASTM (C-425-64 of the latest edition).

5-7 What are the fusing requirements for burned lead joints?

Lead burned (welded) joints shall be lapped and fused together to a thickness of at least that of the material being used.

5-8 How are asbestos cement sewer pipe joints made?

Joints in similar diameter *asbestos pipe* shall be made with sleeve couplings of the same material as the pipe and seated with approved gaskets or rings. Joints between *asbestos cement pipe* and metal pipe shall be made by means of a proper adaptor coupling caulked in a manner suggested in Question 5-1 of this section. Asbestos pipe may no longer be permitted. Check local code.

5-9 Are fiber pipe bituminized joints permitted for all installations?

It is not recommended that fiber pipe bituminized joints be used where temperatures exceed readings of 140°F. At locations other than high temperatures, joints shall be made with tapered couplings of the same material as the pipe.

5-10 Are cold poured joints acceptable for installation below the Earth water table for clay or concrete sewer pipe?

Cold poured resin or bituminous joints are *not* recommended for clay or concrete sewer pipe installed in locations below the Earth water table.

5-11 What are the recommendations for expansion joint materials in piping?

It is recommended that expansion joint materials conform to the type of piping in which the expansion joint is being installed and that all expansion joints be accessible for inspection purposes.

5-12 How shall pipes through walls and roofs be installed?

Pipes through roofs and exterior walls shall be made watertight with proper and approved flashings. Figure 5-4 illustrates a typical lead roof flashing.

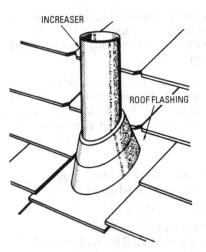

INCREASER

ROOF FLASHING

Figure 5-4 Typical lead vent roof flashing.

5-13 Are increasers or reducers required when connecting pipe and fittings of different sizes?

Where pipe and fittings of different sizes are to be joined, proper reducing fittings or increasers are used. Figure 5-5 illustrates typical pipe increasers and reducers.

(A) Bell-and-flange cast iron reducer.

(B) Steel pipe screw joint reducer.

(C) Vitrified clay pipe increaser.

Figure 5-5 Pipe increasers and reducers.

5-14 Are there prohibited joints and connections?

Any fitting or connection that has a chamber, enlargement, or recess with a ledge shoulder or reduction of the pipe area in the direction of flow, and that would interfere or retard the flow on the outlet or drain side of any trap, is prohibited.

5-15 What is meant by *rough-in*?

Rough-in means the installation of certain parts of the plumbing system that can be done before installation of the fixtures (including drainage, venting, and water supply).

5-16 What is a sewer pipe saddle?

A *saddle* is a correctly contoured device that is commonly used in gravity sewer systems for each variety of pipe. Saddles are specifically made in sizes for lateral pipe connections to main sewer lines (see Figure 5-6).

Figure 5-6 Typical sewer pipe saddle. *(Courtesy General Engineering Co.)*

5-17 What are some of the recommendations for establishing leak-free gasket placement at bolted pipe joints?

Use a gasket material that conforms easily to flange surfaces and lay the gasket as close as possible to the inside of the bolts to help reduce flange distortion. If a fibrous gasket material such as asbestos is used, be sure the gasket is wide enough to seal all leakage paths. *Be sure flange surfaces are clean before using gasket material.* In Figure 5-7, a "marshmallowy" rope-type gasket sealant that has high compressibility has been used.

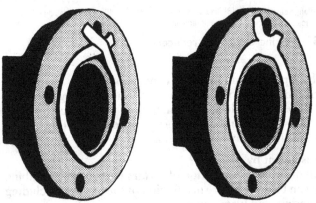

Figure 5-7 Placing a soft gasket material at bolted pipe joints.
(Courtesy W. L. Gore & Associates, Inc.)

True-False Exercises

In the blank to the left of the statement number, place an F if the statement is false or a T if the statement is true. (See Appendix C for answers.)

_____ **1.** The use of jute, hemp, or oakum is not permitted for caulking or for joining water distribution piping connections.

_____ **2.** All plumbing codes require that joints and connections in the plumbing system be watertight or gas-tight.

_____ **3.** Joints in lead pipe (or between lead pipe or fittings) and brass or copper pipe, ferrules, soldering nipples, bushings, or traps shall be (in some instances) full-wiped joints.

_____ **4.** Flared joints are used when making connections in plastic water piping.

_____ **5.** ASTM stands for American Standards for Testing Materials.

_____ **6.** Precast pipe joints are connected with collars.

_____ **7.** Resilient-type materials are used for connecting lengths of plastic pipe in building sewers on both spigot and bell ends.

_____ **8.** It is recommended that fiber pipe bituminized joints be used where temperatures exceed readings of 140°F.

_____ **9.** It is recommended that expansion joint materials conform to the type of piping in which the expansion joint is being installed and that all expansive joints be accessible for inspection purposes.

_____ **10.** A saddle is a correctly contoured device that is commonly used in gravity sewer systems for each variety of pipe.

Chapter 6

Traps, Cleanouts, and Backwater Valves

One of the most important devices in the plumbing system is the *trap*. The trap acts as a form of check valve to permit the free discharge of water or soil from a fixture into the drainage system and prevent the escape of sewer gas into the fixture and room where the fixture or fixtures are located. Figure 6-1 illustrates a typical P-trap fixture. Figure 6-2 illustrates a typical deep-seal cast-iron drainage trap.

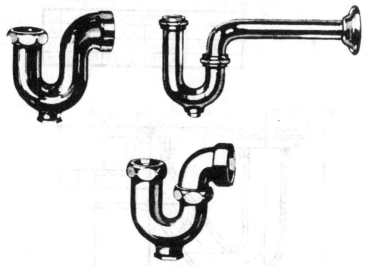

Figure 6-1 Various P-trap fixtures.

The following are typical questions asked in reference to traps, cleanouts, and backwater valves in plumbing licensing examinations.

6-1 At what locations are plumbing traps required?

Each plumbing fixture (except those with integrated traps) is required to be separately trapped by a water-seal-type trap placed close to the fixture. A combination fixture with up to two laundry trays and one sink, or three laundry trays, may connect with a single trap if the trap is placed centrally and the branches connect into the trap seal. Waste outlets should not be more than 30 inches apart

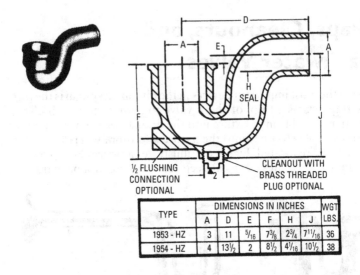

TYPE	DIMENSIONS IN INCHES						WGT
	A	D	E	F	H	J	LBS.
1953 - HZ	3	11	$^5/_{16}$	$7^3/_8$	$2^3/_4$	$7^{11}/_{16}$	36
1954 - HZ	4	$13^1/_2$	2	$8^1/_2$	$4^1/_{16}$	$10^1/_2$	38

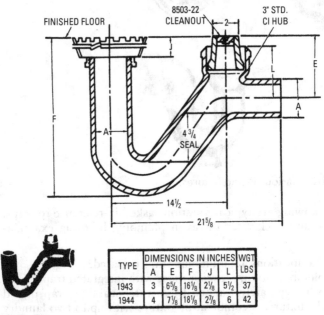

TYPE	DIMENSIONS IN INCHES					WGT
	A	E	F	J	L	LBS
1943	3	$6^5/_8$	$16^1/_8$	$2^1/_8$	$5^1/_2$	37
1944	4	$7^1/_8$	$18^1/_8$	$2^7/_8$	6	42

Figure 6-2 A typical deep-seal cast-iron drainage trap.
(Courtesy Josam Mfg. Co.)

and compartments not more than 6 inches deeper than the others. No fixture shall be double-trapped.

6-2 Is there a relationship between trap and drain size?

Traps shall be of the same size as the fixture drain to which the trap connects. Every trap shall be of the self-cleaning type. Normal floating solids or sediments shall be removed by liquid discharge from the connected fixture.

6-3 Is a water seal required for traps?

Each fixture trap shall have a water seal of not less than 2 inches and not more than 4 inches except where a deeper seal is required for special conditions by administrative authority.

6-4 What is the purpose of a trap seal primer and how is it used?

A *trap seal primer* (see Figure 6-3) is used for installation on cold water supply lines to kitchen sinks, toilets, drinking fountains, and similar fixtures, with connection to traps on infrequently used waste lines for purposes of maintaining a trap seal.

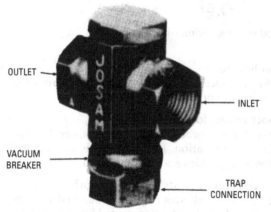

OUTLET

INLET

VACUUM
BREAKER

TRAP
CONNECTION

Figure 6-3 A trap seal primer. *(Courtesy Josam Mfg. Co.)*

Frequently this water seal will evaporate from the trap, thus allowing sewer gas and vermin to escape into a home or building from the sewer lines.

A trap seal primer valve (see Figure 6-3), when properly installed in a cold-water supply line and connected to a floor drain trap, will ensure a constant seal of fresh water in the drain trap.

This valve operates on the principle that, as water is drawn through the supply line by an appliance or water closet, the resulting flow activates the valve mechanism that meters a precise amount of water from the supply line into the trap line. Figure 6-4 shows a typical trap seal primer installation in a water supply line. The vacuum breaker on the primer valve positioned directly over the drain trap connection prevents siphonage of trap water into the water supply line.

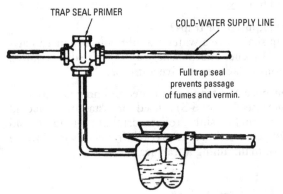

Figure 6-4 A typical trap seal primer installation.

6-5 Are any traps prohibited?

Traps that depend on concealed interior partitions or movable parts for action are generally prohibited for fixture use.

6-6 What is the proper setting for traps?

All traps shall be set level to protect their water seals and to protect the water seal from evaporation. Freeze protection is required in locations where freezing problems are encountered.

6-7 Are there any special requirements for trap design?

Traps for fixtures shall be self-scouring without interior partitions. Fixtures with integral traps may be permitted by some plumbing administrative authorities, but only if such traps have a uniform smooth waterway.

6-8 Is trap siphonage permitted?

Fixture trap seals shall be protected so as to prevent siphonage or back pressure by providing air circulation with an approved vent. No vent shall be connected at the crown of a trap.

6-9 What is the proper support for an acid-resisting type underground trap?

In underground locations where acid-resisting traps (such as vitrified clay or similar type brittle materials) are installed, traps shall be imbedded in concrete extending a minimum of 6 inches at the bottom and sides.

6-10 Are there pipe cleanout size requirements?

Cleanouts shall be of a size corresponding to the pipe size into which they are installed up to a diameter of 4 inches. Cleanout sizes shall not be less than 4 inches for pipe larger than 4 inches in diameter.

6-11 Where are cleanouts to be located?

Waste pipe from sinks or other similar fixtures shall have enough accessible cleanouts over the entire length to ensure cleaning by rodding. A cleanout shall be located near the junction of the building drain and building sewer, or a wye branch cleanout may be placed inside the building near the wall.

6-12 Are cleanouts required for changes in pipe direction?

At each change of building drain direction greater than 60°, a cleanout shall be installed.

6-13 Is there a preferred way to install cleanouts?

Cleanouts shall be installed so that the cleanout opens in a direction opposite to drainage flow or at a right angle to the flow.

6-14 Is it necessary to install cleanouts in stacks?

Cleanouts shall be provided at locations no more than 3 feet above the vertical waste or soil stack.

6-15 Is it permissible to conceal cleanout plugs?

It is not advisable to conceal cleanouts, but when necessary, a covering plate or access door shall be provided permitting easy access to the cleanout. The following figures illustrate installation diagrams for typical Josam Mfg. Company cleanouts installed in any construction, subject to modification of construction in accordance with on-site conditions and cleanout features. Figures 6-5 through 6-9 show floor cleanout installation diagrams. Figure 6-10 and Figure 6-11 are illustrations of access covers installed when piping is concealed. Figure 6-12 shows a carpet cleanout marker and Figure 6-13 an internal cleanout closure plug.

6-16 What are some recommendations for installing floor drains?

Floor drains are available in a variety of sizes and types. They are generally installed in accordance with the best trade practices and manufacturers' recommendations. Figure 6-14 shows typical

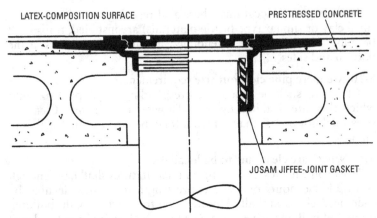

Figure 6-5 A wide flange cleanout for composition surface.
(Courtesy Josam Mfg. Co.)

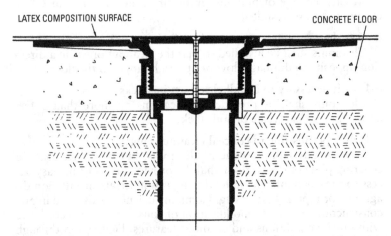

Figure 6-6 An adjustable wide flange cleanout for composition surface. *(Courtesy Josam Mfg. Co.)*

installation diagrams subject to modification of construction in accordance with the drain features.

6-17 When are backwater valves required in drainage systems?
   Backwater valves are required for drainage line installation in locations where fixtures are subject to backpressure or backflow.

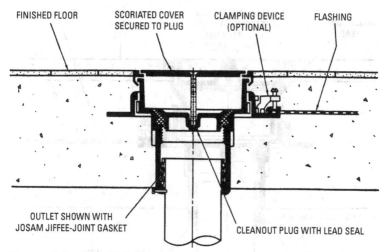

FINISHED FLOOR   SCORIATED COVER   CLAMPING DEVICE   FLASHING
SECURED TO PLUG   (OPTIONAL)

OUTLET SHOWN WITH
JOSAM JIFFEE-JOINT GASKET

CLEANOUT PLUG WITH LEAD SEAL

Figure 6-7   An adjustable floor cleanout. *(Courtesy Josam Mfg. Co.)*

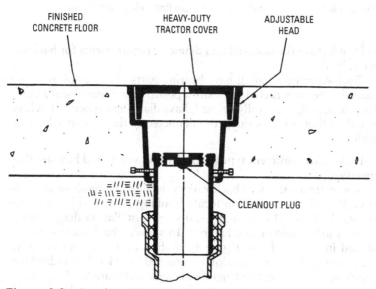

FINISHED   HEAVY-DUTY   ADJUSTABLE
CONCRETE FLOOR   TRACTOR COVER   HEAD

CLEANOUT PLUG

Figure 6-8   An adjustable heavy-duty floor cleanout. *(Courtesy Josam Mfg. Co.)*

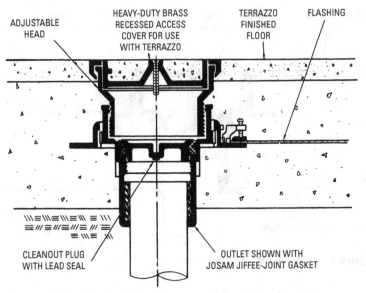

ADJUSTABLE HEAD

HEAVY-DUTY BRASS RECESSED ACCESS COVER FOR USE WITH TERRAZZO

TERRAZZO FINISHED FLOOR

FLASHING

CLEANOUT PLUG WITH LEAD SEAL

OUTLET SHOWN WITH JOSAM JIFFEE-JOINT GASKET

Figure 6-9 An adjustable terrace-top floor cleanout. *(Courtesy Josam Mfg. Co.)*

6-18 What are the material and diameter requirements for backwater valves?

Backwater valves shall have bearing parts of corrosion-resistant material, be constructed to ensure a positive mechanical seal against backflow, and, when fully opened, have diameter capacity of at least equal to (but not less than) the diameter of the pipe in which installed.

6-19 Are there different types of backwater valves and how are they installed?

There are a variety of backwater valves and methods of installation. Procedures depend on local plumbing codes and job requirements. Figures 6-15 through 6-20 show installation diagrams for various job requirements. Figure 6-15 shows a backwater valve installed in a floor drain. Figure 6-16 shows a backwater valve installation with cover extension to required floor level. Standard soil pipe is used and is cut to length and caulked. Figure 6-17 illustrates a backwater valve offset for use in new construction. Figure 6-18 shows an in-line backwater valve installed in an existing drainage line. Figure 6-19 shows a sewer terminal backwater valve. Figure 6-20 illustrates a manhole check valve installed at the terminal end of a drainage line.

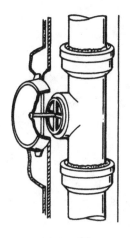

(A) Plaster on metal lath.

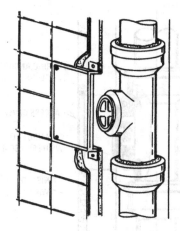

(B) Ceramic tile over plaster board.

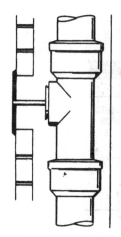

(C) Brick or block wall.

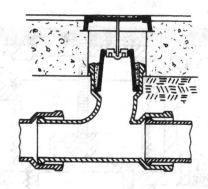

(D) Tile on concrete slab.

Figure 6-10 An access cover installed when piping is concealed.
(Courtesy Josam Mfg. Co.)

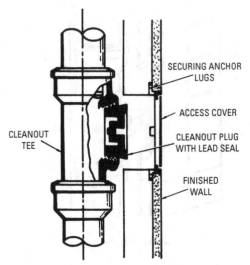

Figure 6-11 A cleanout tee with access cover. *(Courtesy Josam Mfg. Co.)*

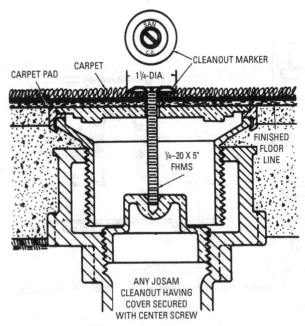

Figure 6-12 A typical carpet cleanout marker. *(Courtesy Josam Mfg. Co.)*

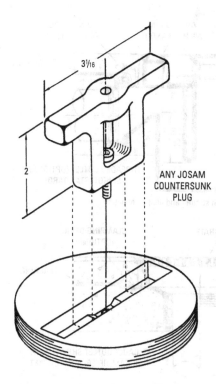

Figure 6-13 A T handle drainage pipe cleanout for internal closure plug.

(Courtesy Josam Mfg. Co.)

3¹/₁₆

2

ANY JOSAM
COUNTERSUNK
PLUG

6-20 What is the purpose of a plumbing trap?

A plumbing trap is a device or fitting used to provide a liquid seal to keep sewer gases out of the house without affecting the flow of wastewater or sewage.

6-21 What does backflow mean when used as a plumbing term?

Backflow means the reverse flow of water or other liquids into water supply pipes carrying potable water.

6-22 What does backflow-preventer mean when used as a plumbing term?

A *backflow-preventer* is a device or means to prevent backflow in piping.

6-23 What is a backwater valve?

A *backwater valve* is a one-way valve installed in the house drain or sewer that prevents flooding of low-level fixtures by backing up of the sewer.

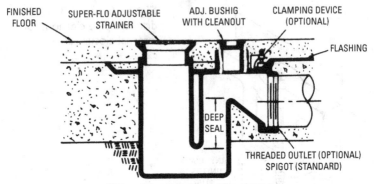

(A) Floor drain with strainer and floor cleanout.

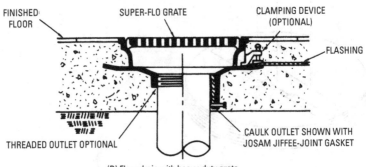

(B) Floor drain with heavy-duty grate.

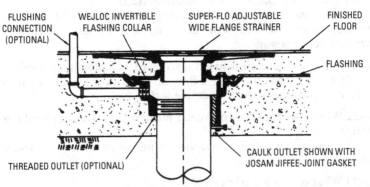

(C) Floor drain with adjustable wide flange strainer.

Figure 6-14 A typical floor drain installation diagram.

(Courtesy Josam Mfg. Co.)

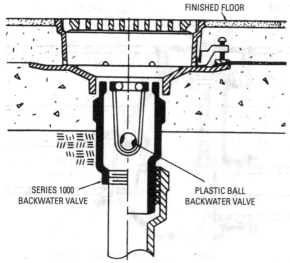

Figure 6-15   A backwater valve installed in a floor drain.
(Courtesy Josam Mfg. Co.)

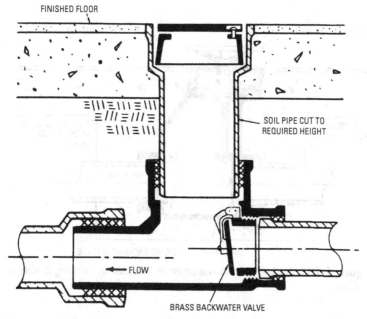

Figure 6-16   A backwater valve installed in soil waste line with extension cover to required level. *(Courtesy Josam Mfg. Co.)*

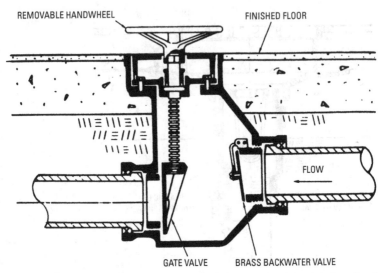

Figure 6-17 A backwater valve, offset for use in new construction.
(Courtesy Josam Mfg. Co.)

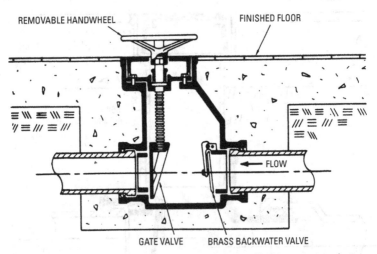

Figure 6-18 An in-line backwater valve installed in existing drainage line. *(Courtesy Josam Mfg. Co.)*

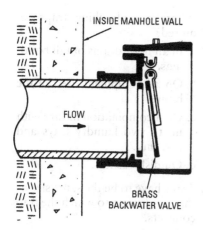

Figure 6-19 A sewer terminal backwater valve installation.

(Courtesy Josam Mfg. Co.)

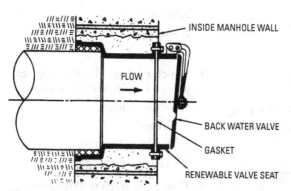

Figure 6-20 A manhole backwater check valve shown installed at terminal end of a drainage line. *(Courtesy Josam Mfg. Co.)*

6-24 What are pipe plugs used for?

Pipe plugs are used to close and seal stub ends of laterals, reserve tees and wyes in sewers, storm drains, manholes, vaults, septic tanks, and conduits. Certain plugs can be used on any type pipe: concrete, asbestos-cement, or clay, or on the spigot end of plastic pipe. Figure 6-21 is a typical pipe plug. Figure 6-22 and Figure 6-23 illustrate the use of pipe plugs in bell and spigot or stub pipe.

Multiple-Choice Exercises

Select the right answer and blacken *a*, *b*, *c*, or *d* on your practice answer sheet. (See Appendix C for answers.)

Figure 6-21 A pipe plug.
(Courtesy General Engineering Co.)

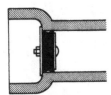

Figure 6-22 A pipe plug used in bell-type pipe.
(Courtesy General Engineering Co.)

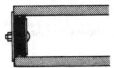

Figure 6-23 A pipe plug used in spigot or stub pipe.
(Courtesy General Engineering Co.)

1. Where are plumbing traps required?
   a. In all floor drains of tall buildings.
   b. On all equipment used in the home.
   c. On a combination fixture with up to two laundry trays and one sink.
   d. On dishwashers.

2. Traps have to be the same size as the _____ drain to which the trap connects.
   a. fixture
   b. floor
   c. sink
   d. wall

3. Which of these is *not* a trap-seal primer required installation?
   a. Kitchen sink
   b. Toilets
   c. Drinking fountains
   d. Garbage disposals

4. Where are traps prohibited?
   a. If they depend on concealed interior partitions or moving parts for action.
   b. If they do not depend on interior partitions or moving parts for action.
   c. If they are over 4 inches in diameter
   d. If they are of the inverted type.

5. Vents cannot be connected _____
   a. at the crown of the trap.
   b. at the end of the trap.

c. at the beginning of the trap.

d. more than 5 feet away.

6. Fixture trap seals shall be protected so as to prevent

   a. siphonage or backpressure.

   b. excessive draining.

   c. free-flowing effluent.

   d. overflows.

7. Cleanouts shall be of a size corresponding to the pipe size into which they are installed up to a diameter of __ inches.

   a. 8

   b. 6

   c. 4

   d. 3

8. Cleanouts are required_____

   a. at every change of direction greater than 60°.

   b. at every change of direction greater than 30°.

   c. at every turn in the pipe.

   d. if the pipe is more than 10 feet long.

9. Backwater valves are required_____

   a. at every fixture location within a multistory building.

   b. in locations where fixtures are subject to backpressure.

   c. in locations where fixtures are arranged in rows.

   d. at free-flowing cold-water pipes in two-story buildings.

10. Pipe plugs are used to seal stub ends of laterals, reserve tees and wyes in _____, storm drains, manholes, vaults, septic tanks, and conduits.

    a. sewers

    b. cold-water lines

    c. hot-water lines

    d. vents

Chapter 7

New Flush Valves, Showers, and Faucets

The following are typical questions asked concerning plumbing fixtures and their accessories in plumbing licensing examinations. Note the newer electronically controlled flush valves made by Sloan Valve Company are shown here. They will be encountered more often as new construction makes use of them.

7-1 What are the qualities required for plumbing fixtures?

All plumbing fixtures shall be made of material with a smooth, impervious surface free from defects and concealed fouling surfaces. However, special sinks and fixtures may be made of chemical stoneware or soapstone. Other special fixtures may be lined and/or made with nickel-copper alloy, lead, or other materials particularly suited for specific purposes.

7-2 What are the special designs or requirements for closet bowls?

Water-flushed closet bowls and traps shall be made in one piece of a form designed to hold enough water (when filled) to the trap overflow to prevent fouling of surfaces when flushed. Integral flushing rims shall be of sufficient size to ensure the flushing of the entire bowl interior.

All Sloan flush valves made since 1906 can be repaired with parts (in kit form) obtained from local plumbing supply houses. It is recommended that the whole inside of the valve be replaced in order to make it like new in operational characteristics. However, it is possible to replace the whole unit with an electronically controlled mechanism and obtain the added convenience of sanitary protection and automatic operation. The Optima Plus uses advanced infrared technology to detect a user's presence and initiate a flushing cycle once the user steps away. The unit is powered by four AA batteries (dry cells) that provide up to three years of service (based on 4000 flushes per month). Alternatively, they may be powered by plug-in, step-down 120-volt to 24-volt transformers. The RES models (see Figure 7-1) are used to convert existing Royal and Regal style flushometers to sensor operation. The 8100 series Optima Plus valves (see Figure 7-2) are complete flushometers valves and ideal for new installations. When installing the valve it is important that the flush model matches the requirements of the plumbing fixture (see Table 7-1).

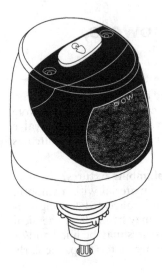

Figure 7-1 Retrofit RESS series—conversion models.

(Courtesy Sloan Valve Co.)

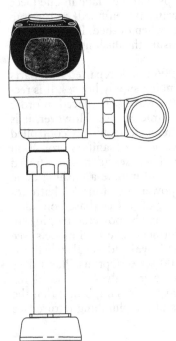

Figure 7-2 Series 8100—complete flushometer model.

(Courtesy Sloan Valve Co.)

Table 7-1 Optima Plus Water Closet and Urinal Models of Flushometers

Optima Plus Water Closet Models	Use
1.6 gpf/6.0 Lpf	For low-consumption bowls
2.4 gpf/9.0 Lpf	For 9-liter European water closets
3.5 gpf/13.2 Lpf	For older water closets
Optima Plus Urinal Models	**Use**
0.5 gpf/1.9 Lpf	For wash-down urinals
1.0 gpf/3.8 Lpf	For low-consumption urinals
1.5 gpf/5.7 Lpf	For older siphon-jet urinals
3.5 gpf/13.2 Lpf	For older blow-out urinals

(*Courtesy Sloan Valve Co.*)

The typical water closet installation shown in Figure 7-3 is the model 8110/8111. When installing the G-2 in a handicap stall, per the Americans with Disabilities Act (ADA) Guidelines (Section 604.9.4), it is recommended that the grab bars be split or shifted to the wide side of the stall. If grab bars must be present over the valve, use the alternate ADA installation shown in Figure 7-4. The alternate ADA installation in Figure 7-4 shows a lower water supply rough-in to 10 inches (254 mm) and a mounted grab bar at the

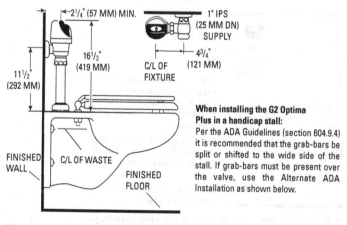

When installing the G2 Optima Plus in a handicap stall:
Per the ADA Guidelines (section 604.9.4) it is recommended that the grab-bars be split or shifted to the wide side of the stall. If grab-bars must be present over the valve, use the Alternate ADA Installation as shown below.

Figure 7-3 Typical water closet installation. *(Courtesy Sloan Valve Co.)*

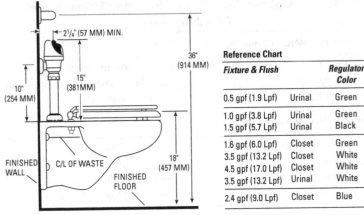

Reference Chart		
Fixture & Flush		**Regulator Color**
0.5 gpf (1.9 Lpf)	Urinal	Green
1.0 gpf (3.8 Lpf)	Urinal	Green
1.5 gpf (5.7 Lpf)	Urinal	Black
1.6 gpf (6.0 Lpf)	Closet	Green
3.5 gpf (13.2 Lpf)	Closet	White
4.5 gpf (17.0 Lpf)	Closet	White
3.5 gpf (13.2 Lpf)	Urinal	White
2.4 gpf (9.0 Lpf)	Closet	Blue

Figure 7-4 Alternate ADA installation. *(Courtesy Sloan Valve Co.)*

36 inches (914 mm) maximum allowed height. The flush volume of the regulator in the flex tube diaphragm kit can be identified by its color.

High rough-in water closet installation (see Figure 7-5) is for models 8113, 8115, and 8116. Models 8115 and 8116 are designed for installation where the water supply is roughed-in 24 inches (610 mm) to 27 inches (686 mm) above the top of the water closet.

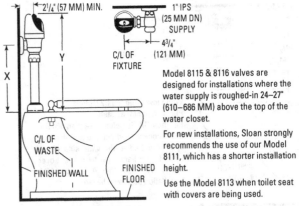

Model 8115 & 8116 valves are designed for installations where the water supply is roughed-in 24–27" (610–686 MM) above the top of the water closet.

For new installations, Sloan strongly recommends the use of our Model 8111, which has a shorter installation height.

Use the Model 8113 when toilet seat with covers are being used.

Figure 7-5 High rough-in water closet installation. *(Courtesy Sloan Valve Co.)*

Table 7-2 Rough-in Dimensions for Various Models of FlushoMeters

Model	X	Y
8113	16 inches (406 mm)	21 inches (533 mm)
8115	24 inches (610 mm)	29 inches (737 mm)
8116	27 inches (686 mm)	32 inches (813 mm)

(*Courtesy Sloan Valve Co.*)

For new installations, Sloan strongly recommends the use of their model 8111, which has a shorter installation height. Use the model 8113 when toilet seats with covers are being used. The x and y measurements for the valve installation are given in Table 7-2. For retrofit, be sure to specify RESS-C for the model being used.

7-3 Are there special types of water closets and seats recommended for public toilets?

Water closets for public use shall be of the elongated type with seats of the open front type made of smooth nonabsorbent materials.

7-4 What is the minimum requirement for water closet flushing procedures?

Water closet tanks are required to have sufficient flushing capacity to properly flush connected water closet bowls. Float valves or ball cocks for low-down flush tanks shall be tight, providing for proper refill of the bowl water trap seal.

Water Closet FlushoMeter

The Sloan exposed, battery-powered sensor-operated G-2 flushometer for floor-mounted or wall-hung top spud bowls come in model numbers 8111 (for low-consumption, 1.6 gpf water closets) and model 8110 (for water-saver, 3.5 gpf models). The control circuit is solid-state and operates on batteries configured to produce 6 volts input to the circuitry. There is an 8-second arming delay and a 3-second flush delay. The units use infrared to operate effectively with a sensor range of 22 to 42 inches, and an adjustable range of ± 8 inches. The power supply is composed of 4 AA batteries, preferably of the alkaline type. This way, the batteries will last about 3 years at the rate of 4,000 flushes per month. Indicator lights are the range adjustment and or low-battery red light (it flashes when the batteries are low). The unit will operate with 15 to 100 psi water pressure. The sentinel flush is set to flush once every 24 hours after the last flush. This keeps the traps filled and the odor from unflushed body wastes from sitting in the bowl for extended periods.

Operation
See Figure 7-6 for the operation of the G-2. Note the positioning of the light beam for effective utilization of the flushing process.

Valve Rough-in
Typical water closet installation can be seen in Figure 7-7A for model 8110/8111. Note the installation caution for the unit in a handicap stall. An alternate ADA installation is shown in Figure 7-7B. Note the lower water supply rough-in at 10 inches, and the grab bar is mounted at the 36 inches maximum allowed height.

7-5 What are the requirements for special wall or flooring materials around public use water closets?
Nonabsorbent flooring materials are required for at least 18 inches from front, and on both sides of the closet, and extending

OPERATION
1. A continuous, invisible light beam is emitted from the OPTIMA *Plus* Sensor

2. As the user enters the beam's effective range (22" to 42") the beam is reflected into the OPTIMA *Plus* Scanner Window and transformed into a low voltage electrical circuit. Once activated, the Output Circuit continues in a hold mode for as long as the user remains within the effective range of the sensor.

3. When the user steps away from the OPTIMA *Plus* Sensor, the circuit waits 3 seconds (to prevent false flushing) then initiates an electrical signal that operates the Solenoid. This initiates the flushing cycle to flush the fixture. The circuit then automatically resets and is ready for the next user.

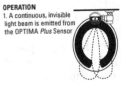
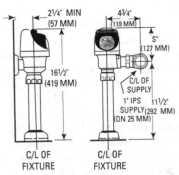

Figure 7-6 Typical water closet installation—Operation.
(Courtesy Sloan Valve Co.)

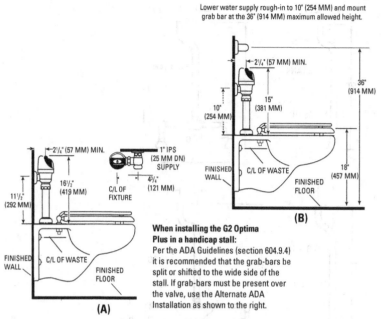

Lower water supply rough-in to 10" (254 MM) and mount grab bar at the 36" (914 MM) maximum allowed height.

When installing the G2 Optima Plus in a handicap stall:
Per the ADA Guidelines (section 604.9.4) it is recommended that the grab-bars be split or shifted to the wide side of the stall. If grab-bars must be present over the valve, use the Alternate ADA Installation as shown to the right.

Figure 7-7 (A) Valve rough-in for water closet installation; (B) Alternate installation for ADA compliance. *(Courtesy Sloan Valve Co.)*

at least 24 inches up the wall behind the public water closet (see Figure 7-8).

7-6 What are the ventilation requirements for toilet rooms?

All toilet rooms or compartments are required to be ventilated either by forced-draft or direct-opening type procedures with the ventilation extending to the atmosphere.

7-7 Are there requirements for properly securing water closet and urinal fixtures?

All wall-hung fixtures shall be rigidly secured by metal hangers or bolts and floor fixtures by properly placed screws or bolts. In addition, fixtures shall be so installed as to afford easy access for cleaning.

7-8 How many water closets and/or urinals can you connect to one flush valve, and what is the operational requirement of such a valve?

Not more than one fixture shall be served by a single flush valve. Each valve at each operation shall provide water in sufficient amount

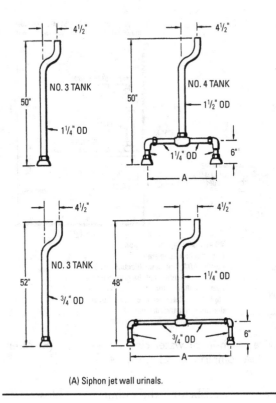

(A) Siphon jet wall urinals.

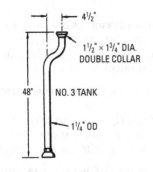

(B) Blowout urinals.

Figure 7-8   Dimensions for flush pipe assemblies. *(Courtesy Sloan Valve Co.)*

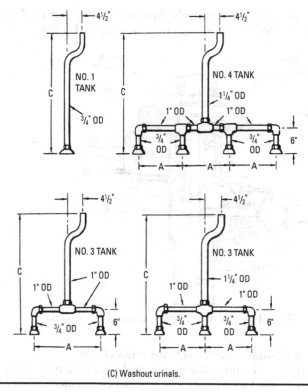

(C) Washout urinals.

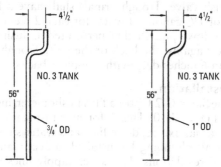

(D) Cast-iron trough urinals.

Figure 7-8 *(Continued)*

at a rate of delivery to completely flush the fixture and refill the fixture trap. Figure 7-9 illustrates a typical toilet flush valve.

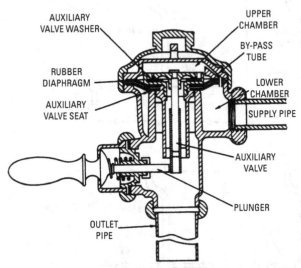

Figure 7-9   Typical type of manual flush valve. *(Courtesy Sloan Valve Co.)*

7-9 Are there flushing requirements for urinals?

All urinals shall be provided with a flushing capacity sufficient for the type of urinal used either by a hand-operated mechanism or by an automatic flush valve. Trough urinals shall have a flushing minimum capacity of $1\frac{1}{2}$ gallons of water for each 2 feet of urinal trough length. Wash-down pipe shall be perforated to flush with an even curtain of water against the back of the unit. Trough urinals shall be not less than 6 inches deep with one-piece backs.

Typical Urinal Installation

Prior to installing the Sloan G-2 Optima Plus flushometer, install the items illustrated in Figure 7-10. This is for new installations only. Install the closet or urinal fixture, drain line, and water supply line. It is important that all plumbing be installed in accordance with applicable codes and regulations. The water supply lines must be sized to provide an adequate volume of water for each fixture. Flush all water lines prior to making connections. The G-2 is designed to operate with 15 to 100 psi (104 to 680 kPa) of water pressure. The minimum pressure required to the valve is determined by the type

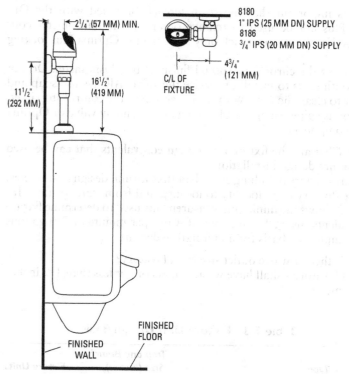

Figure 7-10 Typical urinal installation using models 8180 and 8186.
(*Courtesy Sloan Valve Co.*)

of fixture selected. Consult the fixture manufacturer for pressure requirements. Most low-consumption water closets (1.6 gallons/ 6 liters) require a minimum flowing pressure of 25 psi (172 kPa).

Tools Required for Installation

To adjust the *control stop*, a slotted screwdriver is needed. The Sloan A-50 "Super Wrench" or smooth-jawed spud wrench for couplings will also come in handy. The trim pot adjustment screwdriver is supplied with the unit and is needed to adjust the range in some instances. The strap wrench is also supplied to aid in the installation of the Optima Plus to the valve body. A $\frac{7}{64}$-inch hex wrench is also supplied and will be needed to secure the Optima Plus cover to the base plate.

It is important that the *strap wrench* provided with the Optima Plus not be used to remove or install the flushometer couplings. Use the strap wrench only to install the Optima Plus *locking ring*.

Protect the chrome or special finish of the flushometer. Do not use tooth tools to install or service the valve. Always use soap and water to clean the unit. With the exception of the control stop inlet, do not use pipe sealant or plumbing grease on any valve or Optima Plus component!

7-10 What are the fixture unit design equivalents that can be used for fixture design installation?

Various state plumbing codes have fixture unit design basis tables. These do vary from locality to locality, and from state to state. Table 7-3 shows the minimum measurements used to determine fixture equivalents (using 1 cubic foot of water per minute or $7\frac{1}{2}$ gallons per minute as a basis for a unit fixture design).

7-11 Is there a waste outlet size for a lavatory?

All lavatories shall have waste outlets of not less than $1\frac{1}{4}$ inches in diameter.

Table 7-3   Fixture Unit Design Basis

Fixture Type	Trap and Branch Size (inches)	Fixture Units
Bath (tub, sitz, foot)	$1\frac{1}{2}$	2
Bath (tub and shower combination)	$1\frac{1}{2}$	2
Bath (shower)	2	3
Combination fixture	$1\frac{1}{2}$	3
Drinking fountain	1	$\frac{1}{2}$
Floor drain	2	3
Laundry tray	$1\frac{1}{2}$	3
Lavatory	$1\frac{1}{4}$	1
Sink (residence with dishwasher)	$1\frac{1}{2}$	$1\frac{1}{2}$
Sink (pantry, bar)	$1\frac{1}{4}$	1
Sink (public, hotel)	2	3
Urinal (lip)	$1\frac{1}{2}$	2
Urinal (pedestal)	3	4
Urinal (trough, stall)	2	3
Water closet	3	6

Battery-Operated Faucets

The automatic faucets you may have encountered in commercial and industrial rest rooms are now becoming desirable in homes. It is expected that more of a demand will be evident in the future for automatic devices as the younger generation becomes familiar with the convenience and sanitary advantages of the units. The automatic faucet is but one of these bathroom electronically controlled devices. The sensor and the solenoid are about all that is new and different. The plumber must become more familiar with these in order to service and install them.

Sloan's Optima Plus EBF-85 faucet is shown in Figure 7-11, with the parts detailed in Figure 7-12 and listed in Table 7-4 where they are keyed to and named in the parts list. The current solenoid in production since May 1998 (part 7 shown in Figure 7-12) has a male inlet and a male outlet connection. The current style filter plug has a metal filter cap that can be removed with a $5/8$-inch wrench. The older solenoid valve can be identified by a female inlet and a female outlet connection. These require EBF-17 compression fittings. The older-style solenoid has either a plastic or a replacement brass filter cap. Remove the older brass filter cap with a $7/16$-inch or $1/2$-inch hex wrench. Replace the old style filter cap with an EBF-1005-A filter cap replacement kit that has a brass cap.

Regular maintenance requires the replacement of batteries in the battery-operated model. However, the transformer models do not need this attention. To replace the batteries, loosen the security setscrew on the coupling ring with a 0.05-inch hex or Allen wrench. Unscrew the battery compartment coupling ring. Remove the battery compartment. To ensure proper operation, remove old batteries and insert four new C-size alkaline batteries. Check that the orientation of each battery matches the positive (+) and negative (−) symbols shown in the bottom of the battery compartment. Reattach the battery compartment to the control module by aligning the arrow on the battery compartment with the arrow on the solid tab of the solenoid valve module. Secure by tightening the battery compartment coupling ring. To deter unauthorized removal of batteries, use a 0.05-inch hex wrench to tighten the security setscrew on the coupling ring. See Figure 7-9.

7-12 What type of shower receptacles and shower heads are required for institutional use?

All shower receptacles shall have watertight pans, except those built directly on the ground. When shower receptacles are required for ground surfaces, such receptors shall be built of dense

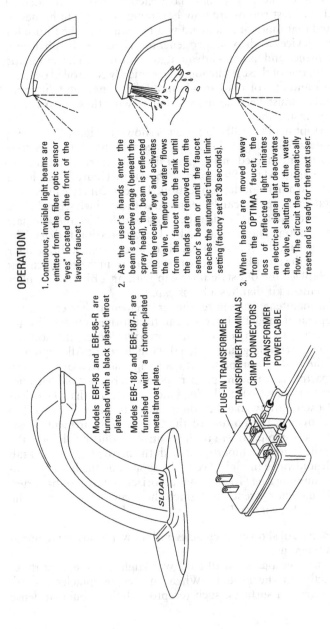

OPERATION

1. Continuous, invisible light beams are emitted from the fiber optic sensor "eyes" located on the front of the lavatory faucet.

2. As the user's hands enter the beam's effective range (beneath the spray head), the beam is reflected into the receiver "eye" and activates the valve. Tempered water flows from the faucet into the sink until the hands are removed from the sensor's beam or until the faucet reaches the automatic time-out limit setting (factory set at 30 seconds).

3. When hands are moved away from the OPTIMA faucet, the loss of reflected light initiates an electrical signal that deactivates the valve, shutting off the water flow. The circuit then automatically resets and is ready for the next user.

Models EBF-85 and EBF-85-R are furnished with a black plastic throat plate.

Models EBF-187 and EBF-187-R are furnished with a chrome-plated metal throat plate.

PLUG-IN TRANSFORMER

TRANSFORMER TERMINALS

CRIMP CONNECTORS

TRANSFORMER POWER CABLE

Figure 7-11 The EBF-85 Optima Plus faucet. *(Courtesy Sloan Valve Co.)*

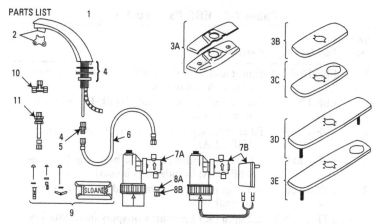

Figure 7-12   The EBF-85 faucet with solenoid. *(Courtesy Sloan Valve Co.)*

nonabsorbent and noncorrosive materials with smooth impervious surfaces.

When shower pans are required for building installation, these pans shall be constructed of at least 4-pound sheet lead or 24-gage copper with corners folded. The corners shall be soldered or brazed in an approved manner and insulated from the rest of the structure with at least 15-pound asphaltic type material.

Act-O-Matic Shower Head

The Sloan Act-O-Matic self-cleaning, wall-mounted shower head with adjustable spray direction is designed for institutional use. The chrome-plated shower head (see Figure 7-13) has the following features:

- It is spring-loaded and self-cleaning with a spray disk that prevents particle clogging and a "cone-within-a-cone" spray pattern for total body coverage.

- It has an adjustable spray angle and pressure compensating (2.5 gpm/9.4 Lpm) flow control and is made of all-brass construction.

- The mounting plate is vandal-resistant with its mounting screws.

- The water inlet is a 1/2-inch IPS pipe nipple with a shower head that is in conformance to all requirements of the ANSI/ASME Standard A 112.18.1M, CSA B-125 and the United States Federal Energy Policy Act.

Table 7-4 EBF-85 Parts List

Item No.	Part No.	Description
1	EBF-10-A	Faucet and sensor assembly (EBF-85 with black plastic throat plate).
	EBF-170-A	Faucet and sensor assembly (EBF-187 with CP metal throat plate).
	EBF-179-A	Fiber-optic sensor cable assembly only (replaces EBF-12-A).
	EBF-1009-A	Fiber-optic sensor cable assembly only (replaces EBF-12-A; includes lens holder).
2	ETF-1021-A	0.5 gpm (1.9 Lpm) spray head with key (female thread).
	ETF-1022-A	2.2 gpm (8.3 Lpm) aerator spray head with key (female thread).
	ETF-237	2.2 gpm (8.3 Lpm) Laminar flow spray head (female thread).
	ETF-435	Replacement key *only* for ETF-1021-A 0.5 gpm (1.9 Lpm) spray head and ETF-1022-A 2.2 gpm (8.3 Lpm) aerator spray head (*not* required for ETF-237 2.2 gpm/8.3 Lpm Laminar flow spray head).
3A	ETF-295-A	4-inch (102 mm) centerset trim plate kit for EBF-85 or EBF-187 faucet (faucet only); includes two nuts, two hex screws, two flat plate washers, single-hole CP cover plate, and black plastic base plate.
3B	ETF-312-A	4-inch (102 mm) centerset trim plate kit for EBF-85 or EBF-187 faucet (faucet only); includes backup spacer, self-tapping screw, and CP single-hole trim plate assembly.
3C	MIX-101-A	4-inch (102 mm) centerset trim plate kit for EBF-85 or EBF-187 faucet with optional mixing valve; includes backup spacer, self-tapping screw, and CP dual-hole trim plate assembly.
3D	ETF-510-A	8-inch (203 mm) centerset trim plate kit for EBF-85 or EBF-187 faucet (faucet only); includes two finger nuts, two washer gaskets, two fender washers, base gasket, backup spacer, and CP single-hole trim plate assembly.
3E	MIX-104-A	8-inch (203 mm) centerset trim plate kit for EBF-85 or EBF-187 faucet with optional mixing valve; includes backup spacer, self-tapping screw, and CP dual-hole trim plate assembly.

(Continued)

Table 7-4 (*continued*)

Item No.	Part No.	Description
4	ETF-290-A	Faucet mounting kit; includes rubber gasket, mounting washer, mounting nut, and ETF-297 compression fitting connector.
5	ETF-297	$1/4$-inch to $3/8$-inch compression fitting connector.
6	MIX-19	20-inch (508 mm) flex hose.
7A	EBF-11-A	Solenoid valve module assembly for battery-powered model; includes solenoid enclosure, solenoid body, and filter cap.
	EBF-4	Four C-size alkaline batteries.
7B	EBF-175-A	Solenoid valve module assembly for transformer-powered model (-R variation); includes solenoid enclosure, solenoid body, and filter cap.
	EBF-176	6-volt power pack for transformer-powered model (-R variation); includes 10-foot (3.05 m) transformer power cable with two crimp connectors.
	EL-386	6-volt transformer for transformer-powered model (-R variation).
8A	ETF-208	$3/8$-inch (10 mm) ferrule.
8B	ETF-209	$3/8$-inch (10 mm) compression nut.
9	EBF-25-A	Mounting bracket kit; includes base plate, wall bracket base, wall bracket, mounting screw, self-tapping screw, two wood screws plus washers and plastic anchors, two screws plus washers and toggle nuts, two screws plus washers and hollow wall anchors.
10	ETF-259	$3/8$-inch (10 mm) tee compression fitting.
11	ETF-470-A	Back check.
—	EBF-1004-A	Solenoid filter replacement kit; includes filter screen assembly and O-ring.
—	EBF-1008-A	Solenoid valve module assembly replacement kit for transformer-powered model (-R variation); includes 6-volt power pack and 6-volt transformer.

Note: Information contained in this table is subject to change without notice. (*Courtesy Sloan Valve Co.*)

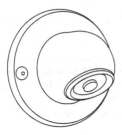

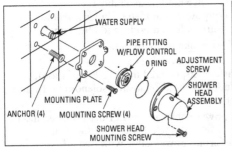

PARTS LIST

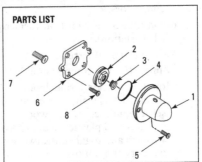

Item No.	Part No.	Description
1	AC-450	Shower Head Assembly
2 *	SH-511	Pipe Fitting
3 *	SH-1005-A	Flow Control, 2.5 gpm (9.4 Lpm)
4 *	SH-512	O Ring
5	SH-514	Shower-Head Mounting Screw (2)
6	SH-510	Mounting Plate
7	K-57	Anchor (4)
8	SH-37	Mounting Screw (4)

* Items 2, 3, and 4 are supplied assembled.

Figure 7-13 The self-cleaning, wall-mounted shower head with adjustable spray direction and integral 2.5 gpm flow control.
(Courtesy Sloan Valve Co.)

Prior to installation, the plumber should be aware of a few things. The shower enclosure should be installed with a drain line and water supply line [½-inch IPS (13 mm DN) male thread required for connection to the shower head].

Important
All plumbing is to be installed in accordance with applicable codes and regulations. Flush all water lines until the water is clear before installing the shower head. The water lines must be sized to provide an adequate flow of water for each shower head. Minimum water pressure of 15 psi (103 kPa) is required.

7-13 What are the special material requirements for public drinking fountains?
Fountains installed for public use shall be made of impervious materials such as enameled cast iron, vitreous china, stoneware, porcelain, or stainless steel.

7-14 How shall backflow *preventers* be installed for public drinking fountains?
Backflow *preventers* shall be installed between the fixture and control valve so as not to be subjected to water pressure except that necessitated by the water flowing to the fixture. Backflow preventers shall *not* be installed on the inlet side of a valve (see Figure 7-14).

7-15 What are the suggested heights for installing the jet nozzle on the fountain?
The jet nozzle and every opening in the pipe must be placed above the edge of the drinking fountain bowl so that possible flooding of the bowl will not result in drain water entering the nozzle of the supply pipe opening.

7-16 Why are public drinking fountains required to have nozzle guards?
Guards of impervious materials shall be provided for the fountain nozzles to prevent the nose and mouth of the user from contacting the nozzle. The nozzle water jet shall be regulated to prevent water spatter.

7-17 Are there special requirements for drinking fountain bowl construction? What are the setting heights for drinking fountains?
The bowls of public drinking fountains shall be designed to be free of sharp corners and smoothly finished to prevent dirt collection and unnecessary splashing. The installation height at the drinking level shall be that height which is most convenient for the greatest majority of users.

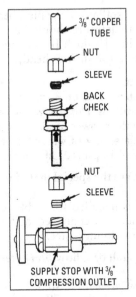

Note: Failure to install the back checks can result in a cross-flow connection when the faucet is in the off position and the supply stops are open. If the pressures of the hot water supply and cold water supply are different, hot water can migrate into the cold water supply or cold water can migrate into the hot water supply. Most plumbing codes require that the back checks be used to prevent this.

Figure 7-14   Backflow preventer. *(Courtesy Sloan Valve Co.)*

7-18 Why are there recommended regulations concerning drain and waste connections?

Because it is recommended that the drain be trapped. If the fixture is not trapped, the drain shall not have direct physical connection with the waste piping.

7-19 What are the recommended regulations concerning openings for public drinking fountains?

The waste opening in the fountain shall be provided with a drain.

7-20 Define slip joint, fixture supply, flood level rim, and flush valve.

A *slip joint* is used primarily on the fixture side of a trap made tight with a rubber plastic-type washer and a slip nut.

A *fixture supply* is a water supply pipe that connects a fixture to a branch supply pipe or to a main water supply pipe.

A *flood level rim* is the edge of a fixture receptacle from which water overflows.

A *flush valve* is a device usually found at the bottom of a toilet tank for purposes of flushing it, or it may resemble Figure 7-9 if it is on a toilet that does not use a tank.

Multiple-Choice Exercises

Select the right answer and blacken *a*, *b*, *c*, or *d* on your practice answer sheet. (See Appendix C for answers.)

1. How many 1.5-volt AA batteries are needed for the flushometer?
   a. 2
   b. 3
   c. 4
   d. 6

2. How many gallons per flush do most low-consumption toilets need?
   a. 1.6
   b. 3.5
   c. 2.0
   d. 5.0

3. What do you use the strap wrench for?
   a. To remove the flushometer couplings
   b. To install the flushometer couplings
   c. To install the Optima Plus locking ring
   d. To remove and reinstall the batteries

4. How are the flush volume regulators identified?
   a. By their size
   b. By their shape
   c. By their edges
   d. By their color

5. What is the light beam range in the G-2 Optima Plus when used on water closets?
   a. 15 inches to 30 inches
   b. 22 inches to 42 inches
   c. 10 inches to 20 inches
   d. 32 inches to 52 inches

6. What should you use to clean the chrome surface of the flushometer?
   a. Bab-0

b. Windex

c. Old Dutch cleanser

d. Soap and water

7. In which finish is the flushometer *not* available?

a. Chrome

b. Brushed nickel

c. Silver

d. Gold

8. What is the minimum pressure needed for operation of the Act-O-Matic shower head?

a. 45 psi

b. 35 psi

c. 25 psi

d. 15 psi

9. How many gallons per flush are needed for modern urinals?

a. 1.6 gallons

b. 1.0 gallon

c. 1.2 gallons

d. 5.0 gallons

10. What is the voltage output of the transformer made for the flushometer?

a. 12 volts

b. 24 volts

c. 30 volts

d. 120 volts

Chapter 8

Fixtures

All plumbers are familiar with those plumbing devices called fixtures. Common fixtures include lavatories, bathtubs, showers, toilets, and urinals. The plumber is primarily concerned with installation procedures, as well as with the various carriers and connections. Figures 8-1, 8-2, and 8-3 illustrate three basic drainage, water, and vent rough-in layouts.

1–BATHROOM GROUP
1–KITCHEN SINK
1–AUTO. WASHER

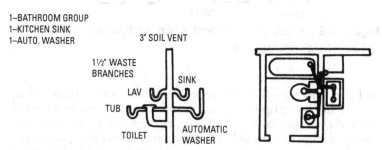

Figure 8-1 A bath and kitchen back to back. *(Courtesy Genova, Inc.)*

1–BATHROOM GROUP
1–WATER CLOSET
1–LAVATORY

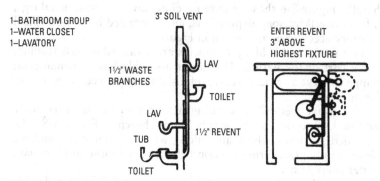

Figure 8-2 A bath up and down. *(Courtesy Genova, Inc.)*

Lavatories

8-1 Are there specific qualities required of lavatories?

The lavatory (by definition, a bowl or basin for washing) is one of the most common fixtures. It is available in many styles, colors,

117

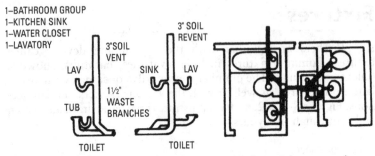

1–BATHROOM GROUP
1–KITCHEN SINK
1–WATER CLOSET
1–LAVATORY

Figure 8-3 A slab job: bathroom group, kitchen sink, water closet, and lavatory. *(Courtesy Genova, Inc.)*

and sizes and is made from pressed steel, cast iron, vitreous china, and plastic. Typical types include the round or oval counter-mounted and self-rimming vitreous china-type. Figure 8-4 is a typical counter-mounted lavatory with all brass supply and indirect lift waste fitting. The diagram shows roughing-in dimensions. Figure 8-5 illustrates a rectangular counter-mounted enameled cast-iron lavatory with front overflow, cast-in soap dishes, and fitting ledge, with a single hooded handle. Figure 8-6 shows a vitreous china lavatory with metal legs, a front overflow, soap depressions, and concealed hangers. Typical lavatory accessories are shown in Figure 8-7.

Hot water and cold water faucets are arranged to pass through some part of the fixture. The faucet assembly has an ornamental flange to serve as a base to give a finished appearance to the installation.

The usual method of securing a faucet to the lavatory body and the connection to the supply line is shown in Figure 8-8. The threaded portion of the shank, along with a coupling nut and the short supply pipe, forms a union between the faucet and the main water supply line.

A typical supply pipe, as shown in Figure 8-9, uses a flexible composition or soft-metal packing in conjunction with a friction ring to provide a leak-proof connection to the faucet.

Waste connection to the lavatory bowl is usually made with a plug top fitting, which is essentially a short length of pipe with a flange at one end and threads at the other end. The two basic types of waste connection (plain and ported) are shown in Figure 8-10. The plain plug top is used for bowls with a separate overflow outlet;

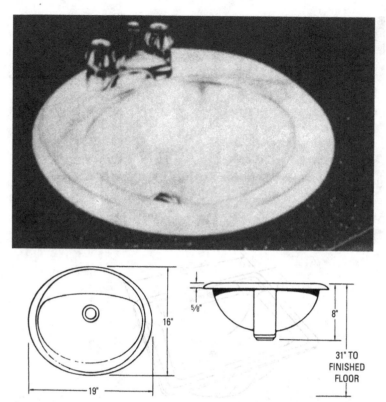

Figure 8-4 An oval, self-rimming counter-mounted lavatory.
(Courtesy Crane Co.)

the ported plug top is used for bowls that have a combined overflow and waste outlet.

Lavatory bowls with an internal overflow passage use a type of waste connection with a pop-up drain as shown in Figure 8-11. The drain stopper is controlled by pushing or pulling a knob located on the faucet assembly.

Bathtubs

8-2 Are there specific qualities required of bathtubs?

Bathtubs are available in various sizes, types, and colors. The most popular type is the rectangular or oblong shaped, and ranges

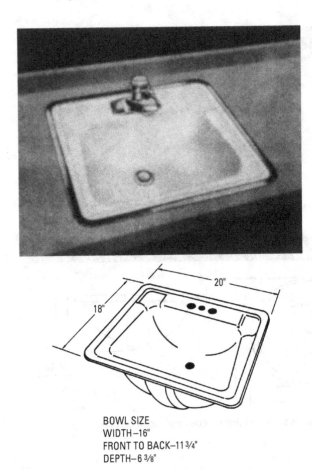

BOWL SIZE
WIDTH–16"
FRONT TO BACK–11 3/4"
DEPTH–6 3/8"

Figure 8-5 A rectangular center-mounted cast-iron lavatory.
(Courtesy American Standard Co.)

in length from 4 to 6 feet. Figure 8-12 illustrates a typical recessed porcelain-enameled cast-iron tub with roughing-in dimensions. Figure 8-13 illustrates a porcelain-enameled cast-iron corner tub with roughing-in dimensions. Figure 8-14 and Figure 8-15 show typical over-the-rim supply fittings complete with bathtub-and-shower combination fixtures with diverter valves to control the flow of water either to the tub or the shower head.

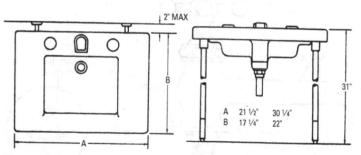

A	21 ½"	30 ¼"
B	17 ¼"	22"

Figure 8-6   A vitreous china lavatory with accessories. *(Courtesy Crane Co.)*

Drain fixtures for bathtubs are available in lever-operated, concealed pop-up bath waste, and overflow with plug-in spud and direct waste outlet adjustable for 14- to 16-inch tubs. They also come in a combination concealed bath waste and overflow with chain and stopper, and are also adjustable for 14- to 16-inch tubs (see Figure 8-16).

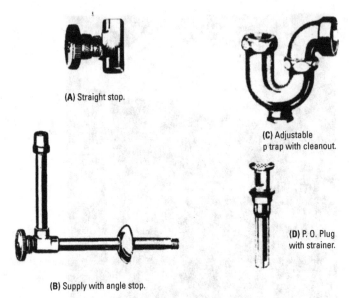

(A) Straight stop.

(C) Adjustable
p trap with cleanout.

(D) P. O. Plug
with strainer.

(B) Supply with angle stop.

Figure 8-7 Typical lavatory accessories. *(Courtesy Crane Co.)*

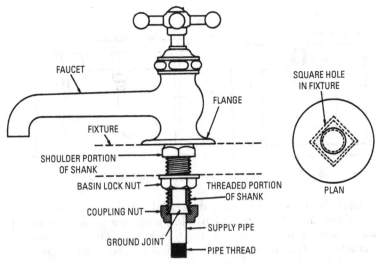

FAUCET

SQUARE HOLE
IN FIXTURE

FLANGE

FIXTURE

SHOULDER PORTION
OF SHANK

PLAN

BASIN LOCK NUT

THREADED PORTION
OF SHANK

COUPLING NUT

SUPPLY PIPE

GROUND JOINT

PIPE THREAD

Figure 8-8 Typical lavatory faucet accessories.

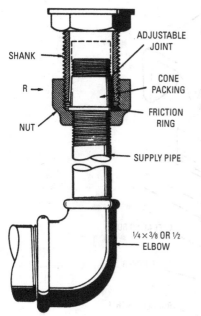

SHANK

R →

NUT

ADJUSTABLE JOINT

CONE PACKING

FRICTION RING

SUPPLY PIPE

¼ × ⅜ OR ½ ELBOW

Figure 8-9 A supply pipe with a cone-type packing.

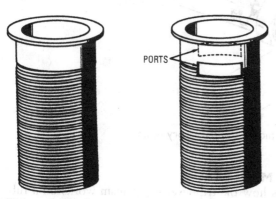

PORTS

Figure 8-10 Lavatory plug top or waste-connection fittings. A plain fitting is shown at the left and a ported fitting is shown at the right.

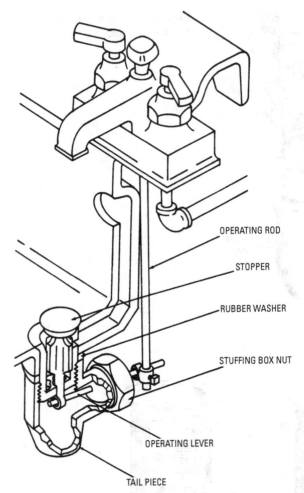

OPERATING ROD

STOPPER

RUBBER WASHER

STUFFING BOX NUT

OPERATING LEVER

TAIL PIECE

Figure 8-11 A typical pop-up lavatory drain.

The Trip-Lever Mechanism

Figure 8-17 shows how the trip-lever mechanism works on a tub drain. To adjust the trip-lever mechanism, remove the screws that hold the overflow plate and pull the entire linkage forward through the hole in the tub. Next, loosen the locknut found at the base of the yoke and then turn the threaded rod clockwise to raise the plunger and counterclockwise to lower the plunger. Then, tighten the locknut

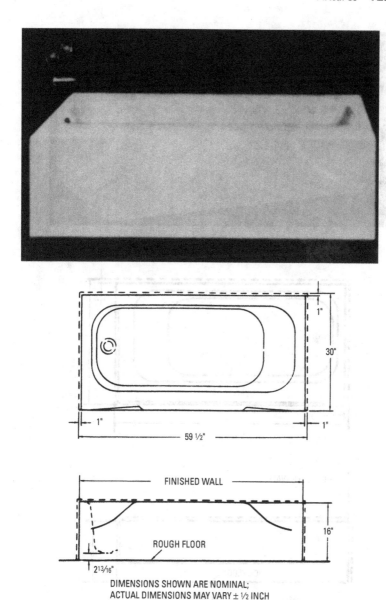

Figure 8-12 A recessed porcelain-enameled cast-iron tub with roughing-in dimensions. *(Courtesy Crane Co.)*

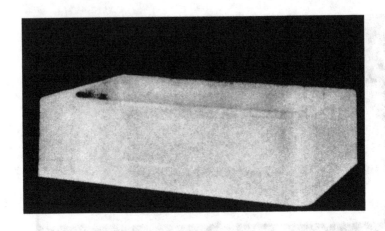

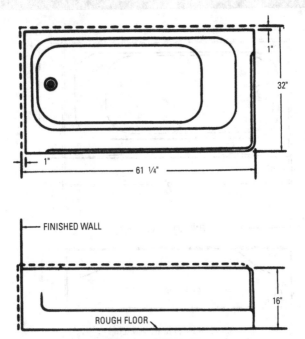

DIMENSIONS SHOWN ARE NOMINAL;
ACTUAL DIMENSIONS MAY VARY + ½ INCH

Figure 8-13 A porcelain-enameled cast-iron corner tub. *(Courtesy Crane Co.)*

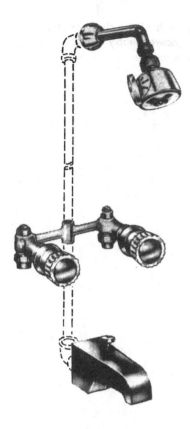

Figure 8-14 An over-the-rim bathtub and shower combination supply connection. *(Courtesy Crane Co.)*

after making the adjustment. Several adjustments may be necessary to get the correct linkage length to ensure that the plunger will seal and open correctly. Too long of a linkage will seal the drain, but not allow proper drainage when opened. Too short of a linkage will prevent proper sealing of the drain and allow water to run out of the tub.

Replace the trip-lever mechanism when it is no longer working properly and cannot be adjusted to stop the flow of water from the tub. Remove the screws while holding the overflow plate and carefully pull the entire linkage out through the overflow opening. The linkage is hinged to allow easy removal. Attach new linkage to the trip lever and slip it into the overflow tube through the tub opening. Lower the linkage carefully until the plunger seats itself properly. It may be necessary to adjust the length of the new linkage

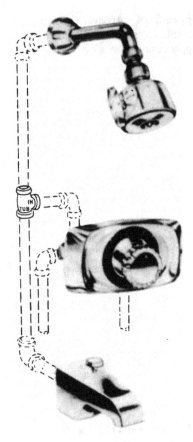

Figure 8-15   A thermostatic tub and shower supply fitting.
(Courtesy Crane Co.)

as described in the preceding paragraph to ensure proper opening and sealing of the drain.

The trip-lever mechanism for pop-up drains is similar to the trip-lever drain (Figure 8-18). However, it ends in a spring rather than a plunger. The spring sits on the rocker linkage and raises or lowers the stopper. Adjust this linkage in the same way as described for the trip-lever mechanism. If repair is indicated for the rocker linkage, lift the stopper and wiggle it sideways, and then gently pull the entire linkage out through the tub drain hole. When you replace the linkage, make sure the bottom curve of the rocker arm faces down.

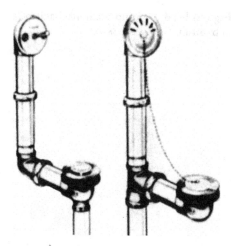

Figure 8-16 Drain fixtures for bathtubs.
(Courtesy Crane Co.)

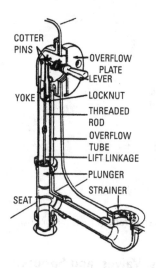

Figure 8-17 Trip-lever mechanism in bathtub. *(Courtesy Plumb Shop.)*

COTTER PINS

OVERFLOW PLATE
LEVER

YOKE
LOCKNUT
THREADED ROD
OVERFLOW TUBE
LIFT LINKAGE
PLUNGER
STRAINER

SEAT

Shower-Tub Diverters

Shower-tub diverters are similar to that shown in Figure 8-19. A diverter functions in the same manner as a faucet. For stem-type diverters, turning the handle causes the stem to move into the valve seat and redirect the water to the shower head.

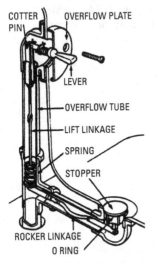

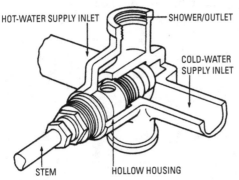

Figure 8-18 Pop-up drain mechanism in bathtub. *(Courtesy Plumb Shop.)*

COTTER PIN

OVERFLOW PLATE

LEVER

OVERFLOW TUBE

LIFT LINKAGE

SPRING

STOPPER

ROCKER LINKAGE

O RING

HOT-WATER SUPPLY INLET

SHOWER/OUTLET

COLD-WATER SUPPLY INLET

STEM

HOLLOW HOUSING

Figure 8-19 Cutaway view of a tub-shower diverter. *(Courtesy Plumb Shop.)*

Repair of Shower and Bath Faucets, Valves, and Spouts
This section reviews how to perform basic repairs on faucets, valves, and spouts.

Repairing the Ball-Type Tub and Shower Faucets
After turning off the water supply valves, drain the lines by turning on the faucet handles. For the lever-style handles, locate the set screw holding the handle and loosen and carefully pry the handle off the stem or use the faucet handle puller.

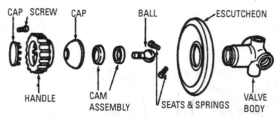

Figure 8-20   Exploded view of tub and shower faucet. *(Courtesy Plumb Shop.)*

Round-style handles first require the removal of the decorative cap (see Figure 8-20), then the removal of the screw. Next, pry off the handle. Cover the cap with tape to protect the finish, and use pliers to unscrew it counterclockwise. Then lift out the cam assembly and ball. Use a pencil or sharp tool to lift out the seats and springs. Check all parts for wear and damage and replace where necessary. On reassembling, be sure that the slot in the ball slips into the pin in the valve body and that the lug on the cam assembly slides into the slot on the valve body. Screw the cap on clockwise and replace the handle. Turn on the water and check for leaks. Tighten the cap further if leaks develop.

Cartridge-Type Shower and Bath Valves

8-3 Are there special recommendations for repairing shower and bath valves?

To repair or replace the shower and bath valves, turn off the water supply and drain the lines by turning on the faucet handle. Pry off the decorative cap and remove the handle screws. Remove the handle and escutcheon. Remove the retainer clip with a pair of pliers and pull the cartridge out of the faucet body (see Figure 8-21). Be sure to note the position of the ears on the cartridge so that they are positioned correctly during reassembly. Check the O-rings and cartridge for wear or damage and replace where necessary. Reverse the procedure for reassembly. Turn on the water and check for leaks.

Shower and Bath Valve Repair

Repairing and replacing the shower and bath valve can be accomplished with the aid of the exploded view shown in Figure 8-22. A faucet puller may be needed if the handle is corroded.

Replacing the Tub Spout

Figure 8-23 shows how the tub spout is removed using a hammer handle. Don't use excessive force, because you could crack the spout.

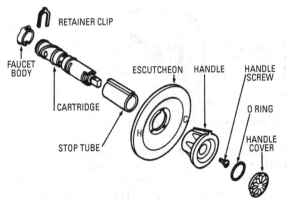

Figure 8-21   Exploded view of cartridge-type shower and bath valves.
(Courtesy Plumb Shop.)

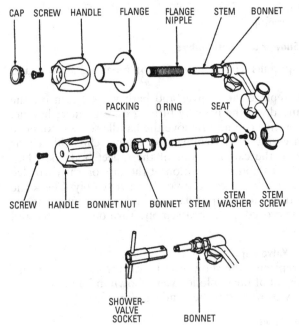

Figure 8-22   Exploded view of shower and bath valve. *(Courtesy Plumb Shop.)*

Do not use Teflon thread, tape, or pipe dope when installing a plastic tub spout.

Shower Head

Figure 8-24 shows the shower head and how it can be disassembled and re-assembled. Be careful not to scratch the chrome finish. Clean the inside of the shower head with a solution of white vinegar and water to remove mineral deposits. Use pipe dope or pipe tape on the shower arm, and screw on the shower head clockwise.

Toilets

8-4 What are the four types of water closet bowls?

The toilet (or water closet) is the most important of all the sanitary fix-

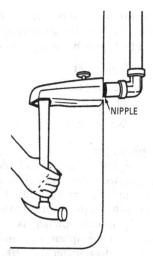

Figure 8-23 Using a hammer handle to remove a tub spout. *(Courtesy Plumb Shop.)*

tures. It is the most complicated of all the fixtures and the least understood. Remember this rule of thumb: The more water you see in the bowl, the better the closet. A large water surface tells you many things: The closet has positive siphon-jet action; it has a strong flushing action; and there is a deep-water seal to guard against obnoxious gases. The toilet's construction, installation, and

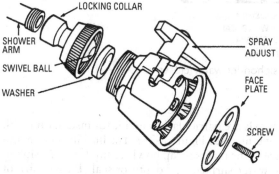

Figure 8-24 Exploded view of shower head. *(Courtesy Plumb Shop.)*

operation are important factors in determining the well-being and health of a building's occupants.

The four general types of toilet bowls are as follows:

- Siphon jet
- Reverse trap
- Blowout
- Washdown

Siphon Jet

The siphon-jet bowl is a logical choice for the most exacting installation (see Figure 8-25). The flushing action of the siphon-jet bowl is accomplished by a jet (*D*) of water being directed through the upleg (*C*) of the trapway. Instantaneously, the trapway fills with water and the siphonic action begins. The powerful, quick, and relatively quiet action of the siphon-jet bowl combined with its large water surface (*A*) and deep water seal (*B*) contribute to its general recognition by sanitation authorities as the premier type of closet bowl.

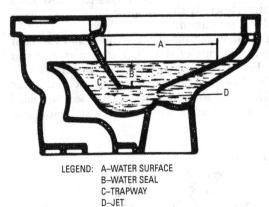

LEGEND: A—WATER SURFACE
          B—WATER SEAL
          C—TRAPWAY
          D—JET

Figure 8-25   A siphon-jet water closet. *(Courtesy Crane Co.)*

Reverse Trap

The reverse-trap bowls are particularly suitable for installation with flush valves or low tanks (see Figure 8-26). The flushing action and general appearance of the reverse-trap bowl are similar to the siphon jet. However, the water surface (*A*), depth of seal (*B*), and size of trapway (*C*) are smaller. Consequently, less water is required for operation.

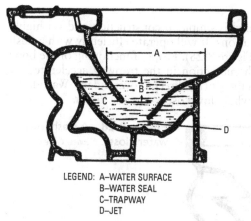

LEGEND: A—WATER SURFACE
B—WATER SEAL
C—TRAPWAY
D—JET

Figure 8-26 The reverse-trap water closet. *(Courtesy Crane Co.)*

Blowout

This type of water closet is operated with flush valves only (see Figure 8-27). The blowout bowl cannot be fairly compared with any other type. It depends entirely upon a driving-jet (*D*) action for its efficiency rather than upon siphonic action in the trapway. It is economical in the use of water, yet has a large water surface (*A*) that reduces fouling space, a deep-water seal (*B*), and a large, unrestricted trapway (*C*). Blowout bowls are especially suitable for use in schools, offices, and public buildings.

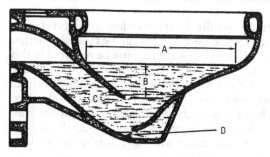

LEGEND: A—WATER SURFACE
B—WATER SEAL
C—TRAPWAY
D—JET

Figure 8-27 The blowout water closet. *(Courtesy Crane Co.)*

Washdown

The washdown-type bowl is simple in construction and yet highly efficient within its limitations. It will operate efficiently with a flush valve or low tank (see Figure 8-28). Proper functioning of the bowl is dependent upon siphonic action in the trapway accelerated by the force of water from the jet (D) directed over a dam. Washdown bowls are widely used where low cost is a prime factor.

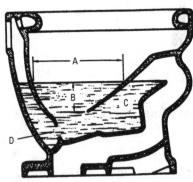

Figure 8-28   A washdown water closet. *(Courtesy Crane Co.)*

LEGEND: A–WATER SURFACE
B–WATER SEAL
C–TRAPWAY
D–JET

Toilet Maintenance

8-5 When is it recommended that a toilet bowl be removed for maintenance?

Stoppages in toilets are usually the result of foreign objects falling into the bowl. Generally, these obstructions can be cleared by either a force cup or an auger. If the obstruction cannot be removed in this manner, it may be necessary to remove the toilet bowl by taking out the bolts holding the bowl to the floor or wall. Figure 8-29 illustrates typical floor-fastened toilet bowl details. A ventilated siphon-jet, elongated-rim, wall-hung bowl with self-draining jet complete with roughing-in details is shown in Figure 8-30. Figure 8-31 shows how a toilet works. A toilet tank cover using a vandal-proof locking device is illustrated in Figure 8-32.

Tanks are used to supply water for flushing in residences and other buildings where quiet flushing is desired or necessary. A minimum amount of water is used during the flushing process. Diaphragm-type flushing valves are used if noise is not a factor. Tank designs

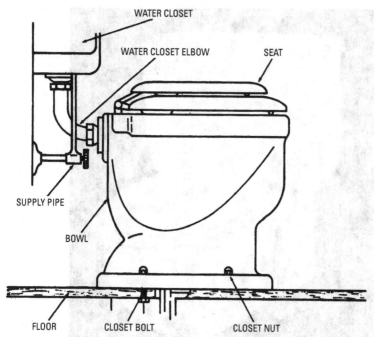

Figure 8-29   Details of a floor-fastened water closet.

may vary, but their mechanisms are similar to those shown in Figure 8-33.

Moisture condensation on the outside surface of the flush tank can be avoided by installing an insulating jacket or securing an insulated flush tank as shown in Figure 8-34. This insulated tank is lined with a monocellular material that is effective in the elimination of tank sweating and dripping. Condensation may also be prevented by a device that mixes a small quantity of hot water with the cold water entering the tank.

Installing a Two-Piece Toilet

8-6 How should a vitreous china, two-piece toilet be installed?

Roughing-in dimensions for a vitreous china close-coupled combination toilet tank and bowl with siphon-jet, whirlpool action, and an elongated rim are shown in Figure 8-35. Installation should be made in accordance with specific on-site conditions, which may vary.

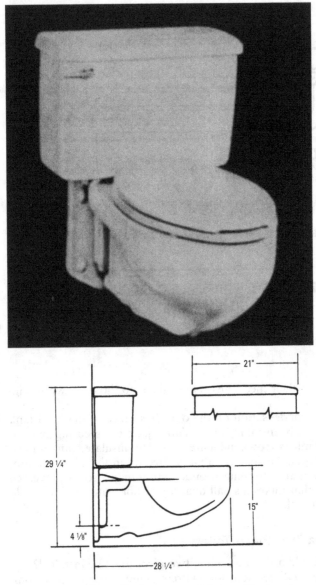

Figure 8-30   Typical siphon-jet, wall-hung toilet bowl. *(Courtesy Crane Co.)*

Water
Rushes
in from
tank
and into
bowl
starting
siphon
action.

Figure 8-31 How a toilet works.
(Courtesy Genova, Inc.)

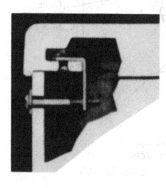

Figure 8-32 A vandal-proof locking device for a toilet tank cover.

To remove the old toilet, follow these steps:

1. Turn off the water supply, and then flush to empty tank. Use a sponge and basin to pick up any water left in the tank or bowl.

2. Disconnect the supply line from the toilet tank.

3. Disassemble the tank from the bowl by removing the locknuts from the underside of the tank.

4. Pry off the bolt caps and remove nuts that secure the toilet to the floor. Use a rocking motion to remove the old toilet.

5. Scrape the old wax gasket from the floor flange.

To install the new toilet, follow these steps:

1. If this is a new bathroom, install a ³/₈-inch IPS water supply and shut-off valve. (Refer to the rough-in drawing for your specific model.)

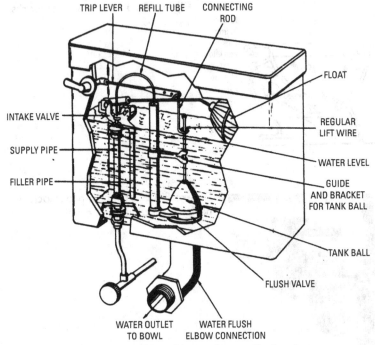

Figure 8-33   A flush-tank mechanism for a toilet bowl.

2. Before installing the new wax gasket, *temporarily* install the flange T-bolts, set the toilet, and place discs and nuts onto bolts. Mark the bolt length needed for bolt cap clearance. Remove the toilet and precut the bolts to length.

3. Carefully turn the toilet bowl upside down onto a thick layer of newspaper to protect the finish. Place a new wax gasket on toilet outlet. A bead of setting compound may be applied to the base if desired (see Figure 8-36).

4. Install two flange T-bolts in the floor flange.

To set the bowl, follow these steps:

1. Carefully lower the toilet so the T-bolts will protrude through the holes in the base. This may require help to align.

2. Apply full body weight to rim of bowl for proper gasket seal.

Figure 8-34   An insulated toilet tank for the purpose of eliminating sweating and dripping. *(Courtesy Crane Co.)*

3. Place the discs (bevel side up) and nuts onto T-bolts and tighten nut (see Figure 8-37). *Do not over-tighten,* because breaking or chipping of china may result. Press the bolt caps onto discs.

To assemble the tank to the bowl, follow these steps:

1. Turn the tank upside down and press the tank to the bowl gasket firmly around the flush valve lockout. Turn tank upright.

2. Slip rubber washers over the two tank bolts (see Figure 8-38). Insert the bolts through the holes in the bottom of the tank and through the tank to bowl gasket.

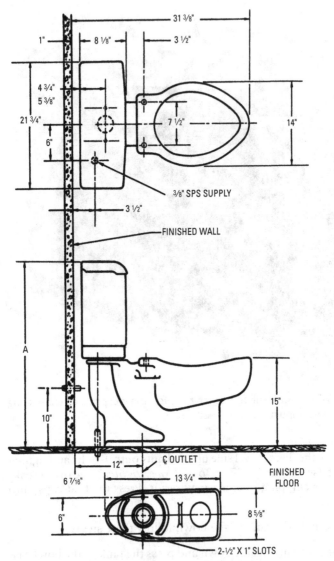

Figure 8-35   Rough-in dimensions for a siphon-jet–type toilet tank and bowl. *(Courtesy Crane Co.)*

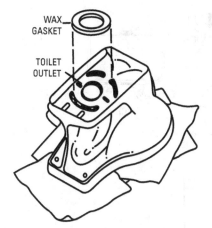

Figure 8-36 Locating wax gasket on 1.6-gallon flush toilet. *(Courtesy of Eljer.)*

3. Lift the tank onto the bowl, carefully aligning the tank bolts with the holes in the bowl. Set the tank in place. From beneath the bowl, assemble the metal washers and nuts onto the tank bolts. Once again, refer to Figure 8-38.

4. Push down on one side of tank and finger-tighten the nut on that side (*A*). Do the same on the other side (*B*).

5. Place a level across top of tank (*C*). Make necessary adjustments to level the tank. Tighten the nuts evenly. *Do not over-tighten,* because breaking or chipping of china may result.

To connect the water, follow these steps:

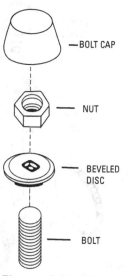

Figure 8-37 T-bolts exploded view. *(Courtesy of Eljer.)*

1. Before connecting the water supply, flush the line into a bucket to remove any debris.

2. Place the coupling nut, friction washer, and universal washer onto the plain end of supply tube, as shown in Figure 8-39.

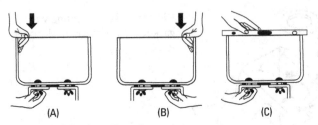

(A)   (B)   (C)

Figure 8-38   Installing the tank. *(Courtesy of Eljer.)*

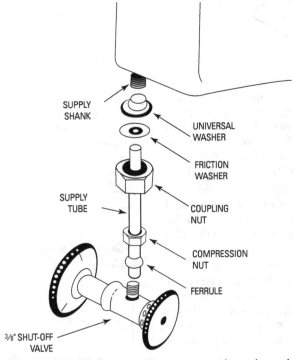

Figure 8-39   Connecting the water supply to the tank. *(Courtesy of Eljer.)*

NOTE

The same universal washer will be used to make the flanged slip-joint connection for a $1/2$-inch OD, or $3/8$-inch flexible, or a $3/8$-inch IPS supply.

3. Temporarily make connection to the tank supply shank, finger-tight. Mark at the shut-off valve the tube length needed and add ¼ inch. (This will insert down into the shut-off valve.)

4. Remove the supply tube and cut to length including the extra ¼ inch. Remove any burrs from inside the cut end. Slip the compression nut and ring (ferrule) onto the shut-off valve end of the tubing.

5. Connect the supply tube to the supply shank once more and finger-tighten. Tubing may be pushed up into tank supply to allow the extra ¼ inch to be inserted into shut-off valve.

6. Tighten both the coupling nut at the tank and the compression nut at the shut-off valve. Make sure the float valve does not rotate while tightening the coupling nut. The float ball must *not* contact the tank walls.

7. Turn on the water supply and the shut-off valve. Check for leaks. An additional tightening adjustment on leaking connections normally will stop the leak.

Making Adjustments

A few minor adjustments may be necessary, including the following:

- *Water level in tank*—Turn the water-level adjustment screw clockwise to lower or counterclockwise to raise the water level to the waterline mark on the tank (see Figure 8-40).

- *Flapper adjustment*—The chain connecting the flush arm to the flapper should have just a little slack so that the flapper will lift completely, yet close fully.

- *Refill tube*—Make sure the refill tube is properly inserted in the top of the flush valve.

Sometimes it is necessary to remove the plastic insert from the refill tube. The insert is easily removed by pushing it through with a small dowel rod.

Before making any adjustments, be sure that the refill tube is properly inserted into the overflow pipe on the flush valve to provide full water seal in the bowl.

Fill the float cup before operating the valve. Turn on the water supply valve. Hold the float cup under water during the first, full fill cycle to fill the bottom half of the cup with water for ballast. The water level can be adjusted by squeezing the water level adjustment

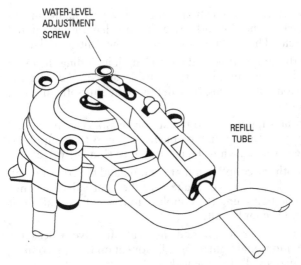

Figure 8-40   Water-level adjustment screw. *(Courtesy of Eljer.)*

clip and moving float cup up or down along stainless steel link (see Figure 8-41).

Be sure that the water level in the tank is always to the water level line marked on the overflow pipe on the flush valve. This is necessary for the unit to consistently operate at the prescribed level of water and maintain satisfactory bowl performance.

If the toilet flushes poorly, the following may be causes:

- Shut-off valve is not fully open.
- Water level in tank is too low.
- Water level in bowl is too low (refill tube blocked or not installed into top of overflow tube).
- Flapper is not lifting and activating flush.

If the float valve does not fully shut off, consider the following:

- Flapper may be defective or warped.
- Parts in top of float valve may be worn.
- Flapper may not be closing completely because of obstruction, or too tight of a chain.
- Float ball may be lodged against tank wall.
- Water level may be too high.

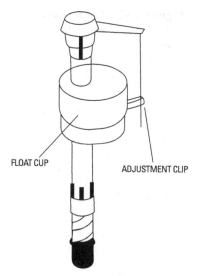

Figure 8-41 Float adjustment clip. *(Courtesy of Eljer.)*

FLOAT CUP

ADJUSTMENT CLIP

The self-adhesive tank cover pads (installed as shown in Figure 8-42) will cushion the contact and prevent rocking.

Care and Cleaning

8-7 How should toilet bowls be cleaned?

Important
Use of any toilet cleaners in the toilet tank void the warranty. In-tank cleaners can damage the rubber, metal, and/or plastic components of tank parts, causing leaks and hindering toilet performance.

To clean bowls, use *toilet bowl cleaners that are applied to the toilet bowl* and a nylon brush when necessary. Use of a metal brush will create metal marks in the bowl.

Urinals

8-8 How should urinals in public restrooms be cleaned?
Urinals receive excessively hard usage because they are normally installed in public restrooms. Considerations in the selection of types should include ease of cleaning and the ability to keep urinals free of debris. A typical exposed-valve type is shown in Figure 8-43 with an installation diagram. This particular model is a vitreous china,

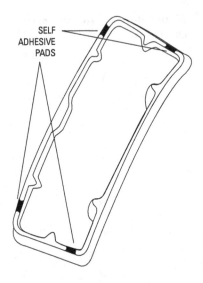

SELF
ADHESIVE
PADS

Figure 8-42   Self adhesive
pads on top of water tank.
(Courtesy of Eljer.)

siphon-jet type with integral extended shields, flushing rim and trap,
$3/4$-inch top spud, 2-inch SPS female outlet connection, and supporting hangers. It easily flushes cigarette butts, gum wrappers, and similar items, thus eliminating the need for strainers and odor cakes.
Siphon-jet action gives blowout efficiency at minimum noise level.

Earthenware and vitreous china are nonabsorbent and are the
most commonly used materials for urinals. There are several types:

- Individual wall type (see Figure 8-43)
- Stall-type with exposed flush valve (see Figure 8-44, including
  the installation diagram)
- Trough-type (see Figure 8-45 for the roughing-in dimensions
  for a trough urinal)
- Pedestal-type (see Figure 8-46).

In basic construction, this urinal is similar to the toilet bowl,
because it is flushed and cleaned by siphon-jet action. Installation
dimensions for a typical siphon-jet action, pedestal-type urinal are
shown in Figure 8-46. Dimensions for flush pipe assemblies for urinals with high water-supply tanks are illustrated in Figure 8-47.
Dimensions will vary depending on specific site conditions.

Fixtures in General
The following are typical questions asked concerning plumbing fixtures and their accessories in plumbing licensing examinations.

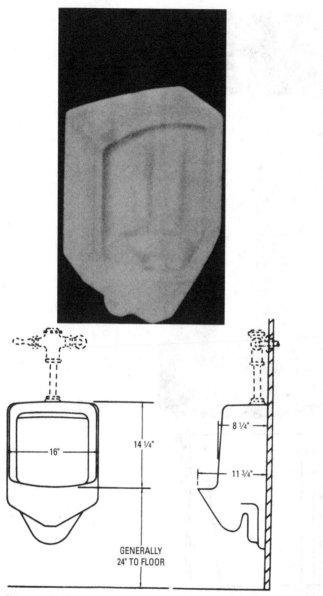

Figure 8-43 A vitreous china siphon-jet urinal with exposed valve.
(Courtesy Crane Co.)

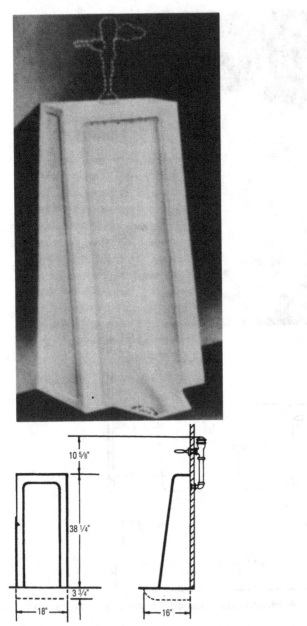

Figure 8-44 A stall-type urinal. *(Courtesy Crane Co.)*

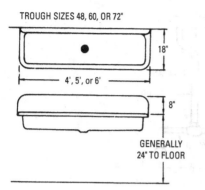

Figure 8-45   Rough-in dimensions for a trough-type urinal.

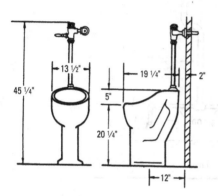

Figure 8-46   A pedestal-type urinal.

8-9 Are specific qualities required for plumbing fixtures?

There are specific qualities required for plumbing fixtures. *All* plumbing fixtures shall be made of material with smooth, impervious surfaces free from defects and concealed fouling surfaces. However, special sinks and fixtures may be made of chemical stoneware or soapstone. Other special fixtures may be lined and/or made with nickel-copper alloy, lead, or other materials particularly suited for specific purposes.

8-10 Are there any special designs or requirements for closet bowls?

Water-flushed closet bowls and traps shall be made in one piece of a form designed to hold enough water (when filled) that the trap overflow can prevent fouling of surfaces when flushed. Integral flushing rims shall be of sufficient size to ensure the flushing of the entire bowl interior.

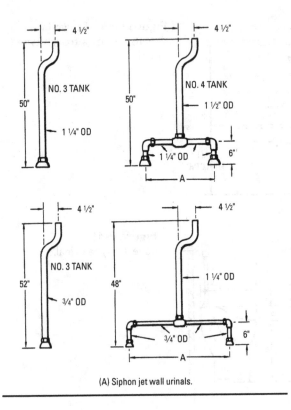

(A) Siphon jet wall urinals.

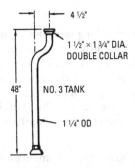

(B) Blowout urinals.

Figure 8-47   Dimensions for flush pipe assemblies. *(Courtesy Crane Co.)*

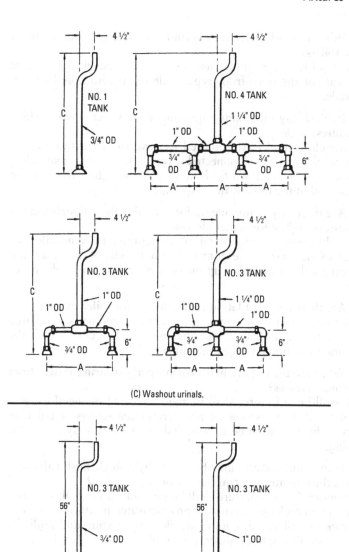

(C) Washout urinals.

(D) Cast-iron trough urinals.

Figure 8-47 *(Continued)*

8-11 What type of water closets and seats are recommended for public toilets?

Water closets for public use shall be of the elongated type with seats of the open-front type made of smooth nonabsorbent materials.

8-12 Is there any minimum requirement for water-closet flushing procedures?

Water-closet tanks are required to have sufficient flushing capacity to properly flush connected water-closet bowls. Float valves or ballcocks for lowdown flush tanks shall be tight, providing for proper refill of the bowl water trap seal.

8-13 Are there any requirements for special wall or flooring materials around public-use water closets?

Nonabsorbent flooring materials are required to be installed for at least 18 inches from the front and on both sides of the closet, and extending at least 24 inches up the wall in back of the public water closet.

8-14 Are there any ventilation requirements for toilet rooms?

All toilet rooms or compartments are required to be ventilated either by forced-draft or direct-opening type procedures with the ventilation extending to the atmosphere.

8-15 What are the requirements for properly securing water-closet and urinal fixtures?

All wall-hung fixtures shall be rigidly secured by metal hangers or bolts and floor fixtures by properly placed screws or bolts. In addition, fixtures shall be so installed as to afford easy access for cleaning.

8-16 How many fixtures may be served by a single flush valve and what is the operational requirement of such a valve?

No more than one fixture shall be served by a single flush valve. Each valve at each operation shall provide water in sufficient amount at a rate of delivery to completely flush the fixture and refill the fixture trap. Figure 8-48 illustrates a typical toilet flush valve.

8-17 What are the flushing requirements for urinals?

All urinals shall be provided with a flushing capacity sufficient for the type of urinal used either by a hand-operated mechanism or by an automatic flush valve. Trough urinals shall have a flushing minimum capacity of $1\frac{1}{2}$ gallons of water for each 2 feet of urinal trough length. Washdown pipe shall be perforated so as to flush

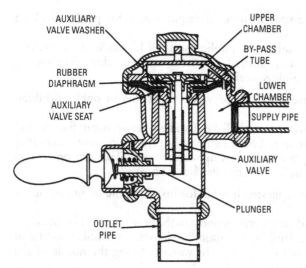

Figure 8-48 A typical toilet flush valve.

with an even curtain of water against the back of the unit. Trough urinals shall be not less than 6 inches deep with one-piece backs.

8-18 Are there any fixture unit design equivalents that can be used for fixture design installation?

Various state plumbing codes have fixture unit design basis tables and these do vary from locality to locality and state to state.

8-19 What is the waste outlet size of a lavatory?

All lavatories shall have waste outlets of not less than $1\frac{1}{4}$ inches in diameter.

8-20 What are the requirements for shower receptacles?

All shower receptacles shall have watertight pans except those built directly on the ground. When shower receptacles are required for ground surfaces, such receptors shall be built of dense, nonabsorbent, and noncorrosive materials with smooth impervious surfaces.

When shower pans are required for building installation, these pans shall be constructed of at least 4-pound sheet lead or 24-gage copper with corners folded. The corners shall be soldered or brazed in an approved manner and insulated from the rest of the structure with at least 15-pound asphaltic-type material.

8-21 Are there special material requirements for public drinking fountains?

Fountains installed for public use shall be made of impervious materials such as enameled cast iron, vitreous china, stoneware, porcelain, or stainless steel.

8-22 Where are backflow preventers installed in public drinking fountains?

Backflow preventers shall be installed between the fixture and control valve so as not to be subjected to water pressure except that necessitated by the water flowing to the fixture. Backflow preventers shall *not* be installed on the inlet side of a valve.

8-23 Are there suggested heights for installing the jet nozzle of the fountain?

The jet nozzle and every opening in the pipe must be placed above the edge of the drinking fountain bowl so that possible flooding of the bowl will not result in drain water entering the nozzle of the supply pipe opening.

8-24 Are public drinking fountains required to have nozzle guards?

Guards of impervious materials shall be provided for the fountain nozzles so as to prevent the nose and mouth of the user from contacting the nozzle. The nozzle water jet shall be so regulated as to prevent water spatter.

8-25 In addition to material requirements, are there any other requirements for drinking-fountain bowl construction? What are the setting heights for drinking fountains?

The bowls of public drinking fountains shall be designed to be free of sharp corners and smoothly finished so as to prevent dirt collection and unnecessary splashing. The installation height at the drinking level shall be the height that is most convenient for the greatest majority of users.

8-26 Are there recommended regulations concerning drain and waste connections and openings for public drinking fountains?

It is recommended that the drain be trapped. If the fixture is not trapped, the drain shall not have direct physical connection with the waste piping. The waste opening in the fountain shall be provided with a drain.

8-27 What is a slip joint?

A *slip joint* is used primarily on the fixture side of a trap made tight with a rubber plastic-type washer and a slip nut.

8-28 What is a fixture supply?

A *fixture supply* is a water supply pipe that connects a fixture to a branch supply pipe or to a main water supply pipe.

8-29 What is a flood level rim?

A *flood level rim* is the edge of a fixture receptacle from which water overflows.

8-30 What is a flush valve?

A *flush valve* is a device usually at the bottom of a toilet tank for purposes of flushing it, or, it may resemble Figure 8-48 if it is on a toilet that does not use a tank.

Multiple-Choice Exercises

Select the right answer and blacken *a*, *b*, *c*, or *d* on your practice answer sheet. (See Appendix C for answers.)

1. Plumbing fixtures are required to have specific qualities. *One* of those qualities is that ____.

    a. they must be made of ceramic material

    b. they must be made of clay

    c. they may be made with smooth, impervious surfaces with freedom of defects

    d. they must be made of stainless steel only

2. Seats for water closets recommended for public toilets ____.

    a. are to have an elongated top with seats of the open-front type made of smooth nonabsorbent materials

    b. are to be round topped with seats of the closed-front type made of smooth nonabsorbent materials

    c. are to be made of stainless steel to prevent transmittal of disease

    d. are to be made of ceramic material to promote self-cleaning

3. Requirements for wall or flooring materials around public-use water closets are ____.

    a. absorbent flooring materials at least 18 inches in front of the fixture

    b. nonabsorbent flooring materials installed at least 18 inches from front and on both sides of the closet and extending at least 24 inches up the wall in back

c. absorbent wall materials over ceramic

d. ceramic materials on walls and floor

4. Requirements for ventilation of toilet rooms include ____.

a. ventilation by fans

b. ventilation by open window

c. ventilation either by forced-draft or direct-opening type extending to the atmosphere

d. ventilation by chemical means to destroy odors

5. All wall-hung fixtures are to be ___ secured by metal hangers or bolts and floor fixtures by properly placed screw or bolts.

a. rigidly

b. firmly

c. sloppily

d. loosely

6. All urinals shall be provided with a flushing capacity sufficient for the type of urinal used either by hand-operated mechanism or by an ____ flush valve.

a. open

b. optical

c. easy

d. automatic

7. All lavatories shall have waste outlets of not less than ___ inch in diameter.

a. 1

b. 2

c. $1\frac{1}{4}$

d. $1\frac{1}{2}$

8. Fountains installed for public use shall be made of impervious materials such as vitreous china, stoneware, _____, or stainless steel.

a. cast iron

b. steel

c. rubber

d. porcelain

9. In public fountains, guards of impervious materials shall be provided for the fountain nozzles so as to prevent the ___ and mouth of the user from contacting the nozzle.

   a. nose

   b. face

   c. ears

   d. chin

10. A slip joint is used primarily on the fixture side of a ___ made tight with a rubber plastic-type washer and a slip nut.

   a. drain

   b. trap

   c. nozzle

   d. faucet

Chapter 9

Interceptors and Special and Indirect Plumbing Wastes

Grease interceptors and special cooking equipment in restaurants, mess-halls, and hospitals (as well as cafeterias) are required by the plumbing code in every state. They are a necessary item in places that use dishwashers commercially or that serve large numbers of diners. Grease produced in dishwashing and cooking equipment that needs clean-up in a sink connected to a plumbing system must be trapped or taken out of the waste water. Grease causes severe damage to the whole plumbing system. Installation and operation of grease interceptors and special cooking equipment are regulated in most municipal areas of the United States.

The following questions and answers are typical in various plumbing licensing examinations.

9-1 Is it necessary to install a grease interceptor for an individual residential unit?

A grease interceptor is not required to be installed for an individual dwelling unit or for private living quarters.

9-2 Where are grease interceptors required and how are they installed?

It is generally required that a grease interceptor be installed on waste lines leading from sinks, drains, or other fixtures of restaurants, hotel kitchens and bars, clubs, factory restaurants or cafeterias, hospital kitchens, or any establishment from which excessive amounts of grease may be introduced into the sewer system. All grease interceptors shall be of a design and so installed that they will *not* allow siphonage or become air bound.

Figures 9-1 and 9-2 show typical grease interceptors and typical capacities with manufacturers' complete detailed information. Acceptable grease interceptors usually carry the approval seal of the Plumbing and Drainage Institute. Figure 9-3 illustrates an installation diagram for a single- or double-compartment sink with grease draw-off installed on the same floor as the fixture being served. Figure 9-4 shows installation for a single-compartment sink with grease draw-off installed on the floor below the fixture being served. Figure 9-5 provides a diagram for a double-compartment sink with the grease interceptor draw-off in a pit below the floor. Figure 9-6 shows an installation diagram for a dishwashing machine with the

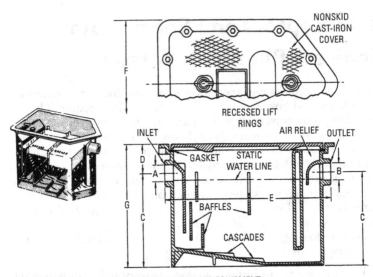

FLOW CONTROL AS LISTED FURNISHED AS STANDARD EQUIPMENT

FLOW RATE GPM	GREASE CAP LBS	INLET A	OUTLET B	ROUGHING CENTERS C	ROUGHING CENTERS D	LENGTH E	WIDTH F	HEIGHT G	WEIGHT LBS	FLOW CONTROL
4	8	2	2	7⅜	3⅛	15⅛	10⅜	10½	50	1041½
7	14	2	2	8⅜	3⅜	16¾	12	11¾	76	1041½
10	20	2	2	9½	3½	18⅞	13½	13	104	1041½
15	30	2	2	11½	3½	21⅝	15¼	15	152	1062
20	40	3	3	13	4⅜	24¾	16⅞	17⅜	200	1063
25	50	3	3	14½	4	26⅜	18	18½	230	1063
35	70	3	3	15½	4⅞	30	19¾	20⅜	300	1063
50	100	3	3	17	6	33⅝	22	23	395	1063

Figure 9-1 Typical grease interceptor complete with cascade bottom, center channel, internal air relief, visible double wall, deep seal, code trap, and removable baffles. *(Courtesy Josam Mfg. Co.)*

grease interceptor recessed flush with the floor. Figure 9-7 illustrates the typical installation of a dishwashing machine, restaurant sink, and dishwasher with the interceptor on the floor with the grease draw-off. Figure 9-8 shows a typical diagram of a double-compartment sink with the grease interceptor installed on the floor. Figure 9-9 illustrates a grease interceptor recessed and serving multiple units on the same floor.

Figures 9-3 through 9-8 carry air intake from the flow control. A local vent or vent stack without a trap can be used if permitted by code, or may terminate under a drainboard with an open end. The

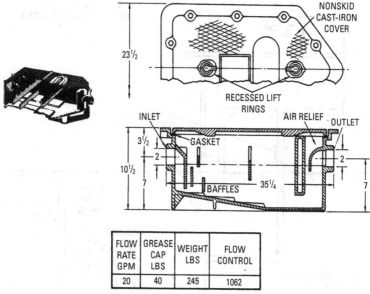

FLOW RATE GPM	GREASE CAP LBS	WEIGHT LBS	FLOW CONTROL
20	40	245	1062

Figure 9-2 A typical cast-iron grease interceptor for installation on the floor. *(Courtesy Josam Mfg. Co.)*

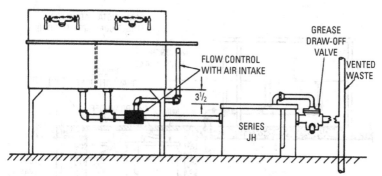

Figure 9-3 Installation for single- or double-compartment sink with grease interceptor having grease draw-off installed on same floor as fixture being served. *(Courtesy Josam Mfg. Co.)*

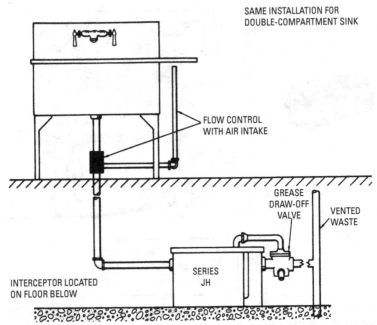

Figure 9-4 Installation designed for a single-compartment sink with grease draw-off installed on the floor below fixture being served.
(Courtesy Josam Mfg. Co.)

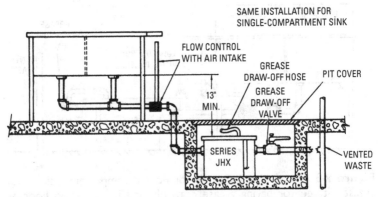

Figure 9-5 Diagram for a double-compartment sink with the grease interceptor draw-off in a pit below the floor. *(Courtesy Josam Mfg. Co.)*

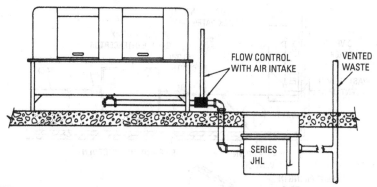

Figure 9-6 Installation for a dishwashing machine with the grease interceptor recessed flush with the floor. *(Courtesy Josam Mfg. Co.)*

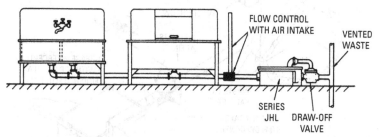

Figure 9-7 An installation of a dishwashing machine and a restaurant sink with the interceptor on the floor with a grease draw-off. *(Courtesy Josam Mfg. Co.)*

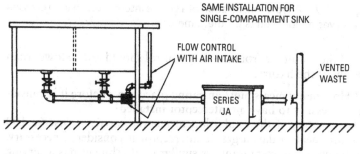

Figure 9-8 Diagram of a double-compartment sink with the grease interceptor installed on the floor. *(Courtesy Josam Mfg. Co.)*

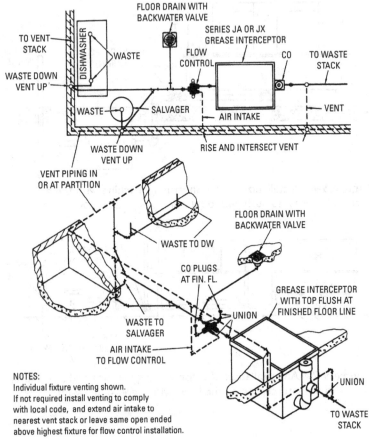

FLOOR DRAIN WITH
BACKWATER VALVE

TO VENT
STACK

WASTE DOWN
VENT UP

DISHWASHER

WASTE

SERIES JA OR JX
GREASE INTERCEPTOR

FLOW
CONTROL

CO

TO WASTE
STACK

WASTE — SALVAGER

AIR INTAKE

VENT

WASTE DOWN
VENT UP

RISE AND INTERSECT VENT

VENT PIPING IN
OR AT PARTITION

FLOOR DRAIN WITH
BACKWATER VALVE

WASTE TO DW

CO PLUGS
AT FIN. FL.

GREASE INTERCEPTOR
WITH TOP FLUSH AT
FINISHED FLOOR LINE

UNION

WASTE TO
SALVAGER

AIR INTAKE
TO FLOW CONTROL

UNION

TO WASTE
STACK

NOTES:
Individual fixture venting shown.
If not required install venting to comply
with local code, and extend air intake to
nearest vent stack or leave same open ended
above highest fixture for flow control installation.

Figure 9-9 A typical installation of a grease interceptor that is recessed and serves multiple units on the same floor. *(Courtesy Josam Mfg. Co.)*

top of the air intake from the flow control should *not* be lower than the rim of the fixture.

9-3 How many meals must an eating place serve before it is considered necessary to have an interceptor installed?

On average, if the establishment has the capacity to serve 100 or more meals a day, a grease interceptor is considered necessary. If a lunch room, restaurant, or similar establishment has a serving or seating capacity of 15 or more patrons at any one time, it is considered to have the capacity to serve 100 or more meals a day.

9-4 What are the preferable locations for installing grease interceptors on drainage lines?

A grease interceptor shall be placed as near as practical to the fixture from which it receives its discharge, and be located outside the building if practical. Protection from freezing is required.

9-5 Must an interceptor be readily accessible for cleaning?

Each interceptor must be readily accessible for cleaning and servicing. Covers shall be easily removable, and when covers are securely attached to the interceptors, they shall be watertight. Cover bolts shall be of noncorrosive material.

9-6 Is there an efficiency requirement for grease interceptors?

Efficiency ratings shall be in accordance with locally accepted standards or practices. However, most local codes accept the following efficiency standard: The flow rate through the interceptor shall not exceed its rated capacity and shall operate at a mean efficiency of at least 90 percent. All codes require the periodic removal of accumulated grease to keep interceptors in an efficiently operating condition and to prevent the entrance of grease or foreign matter into the drainage system.

9-7 If more than one fixture discharges into an interceptor, what are the trapping and venting requirements?

Each such fixture shall be trapped and vented.

9-8 Must interceptors be installed in locations such as garages, beauty shops, and allied establishments?

Codes generally require that interceptors be installed in locations where motor vehicles are serviced and in locations where excessive sediment and/or hair may enter the drainage lines. For motor-servicing locations, design normally requires catch basins of the fixture to be of a size to hold oil, sand, or dirt washings reaching the interceptor during any 10-hour period. In most instances, a basin with a minimum width of 24 inches and a depth below the water level of the basin of at least 24 inches meets requirements. Figure 9-10 shows a typical oil interceptor. Figure 9-11 shows an oil interceptor installed in a drain trench. A sediment or hair interceptor with a removable perforated copper basket having a sloping bottom is illustrated in Figure 9-12. Figure 9-13 illustrates a typical lavatory sediment or hair interceptor.

9-9 Are interceptors required for packing and bottling plants?

Interceptors are required in packing and bottling plants to prevent discharge into the drainage system of materials that could clog

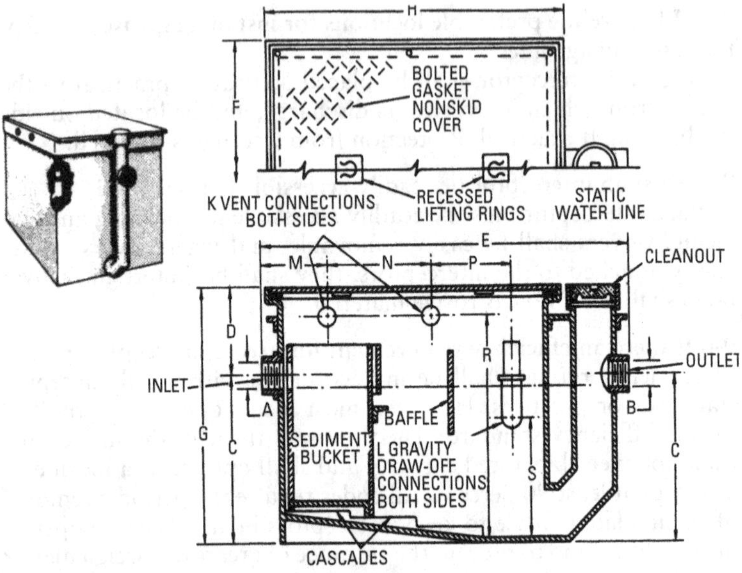

GRC 160 & GRC 170 FURNISHED WITH 2-PIECE COVER.
* FLOW CONTROL AS LISTED FURNISHED AS STANDARD EQUIPMENT.

TYPE	FLOW RATE GPM	DIMENSION IN INCHES																WEIGHT LBS	*FLOW CONTROL
		INLET	OUTLET	ROUGHING CENTERS		LENGTH	WIDTH	HEIGHT	COVER	VENTS	DRAW-OFF	VENT & DRAW-OFF ROUGHING							
		A	B	C	D	E	F	G	H	K	L	M	N	P	R	S			
GRG-110	75	3	3	21	16½	47	31	37½	41	2	2	8¼	12½	10½	34⅛	13½		600	1053
GRG-120	100	4	4	23	15½	54	35	38½	47	2	2	9½	14½	12½	35½	15¼		750	1054
GRG-130	150	4	4	27	16½	60	39	43½	53	2	2	10½	16½	14½	40½	19¼		975	1054
GRG-140	200	5	5	31½	17½	67	42¼	49	58½	2½	2½		18½	16	44¼	26¼		1200	1055
GRG-150	250	5	5	35½	15½	72	46¼	51	63½	2½	2½	12½	20¼	17½	46¼	29¾		1475	1055
GRG-160	350	6	6	39½	15½	80	52¼	55	71½	2½	2½	13½	23	19¾	51⅜	34¼		1875	1056
GRG-170	500	6	6	43½	26½	91	59¼	70	82½	2½	2½	16	27	23	66⅜	38¼		2575	1056

Figure 9-10 An oil interceptor with sediment bucket and information chart for various size units. *(Courtesy Josam Mfg. Co.)*

the drainage system. Figure 9-14 illustrates a typical packing-house grease interceptor with surge control valve and skimming trough.

9-10 What type of plumbing drainage system wastes are considered to be *special*?

There are actually two types of wastes that are considered to be special:

- *Harmful wastes* are wastes considered harmful to the piping of a plumbing system and include acids and alkalines.

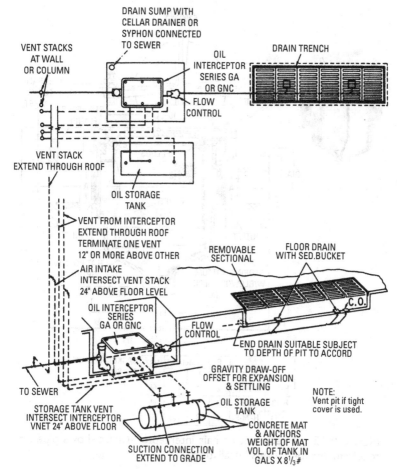

Figure 9-11 An oil interceptor installed in a pit serving a drain trench.
(Courtesy Josam Mfg. Co.)

It is required that these be rendered harmless before being discharged into the drainage system. Pipe, fittings, traps, and connections shall be of materials unaffected by this type of discharge.

- *Prohibited wastes* are wastes that are very toxic, corrosive, explosive, or flammable. Also included in this category are gases, liquids, vapors, or substances that could harm the drainage system or ecology. Occasionally, such substances may be

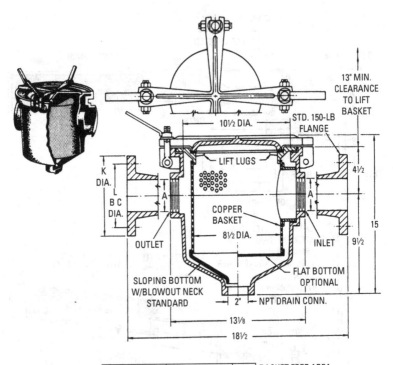

TYPE		Dim. in inches			WGT.	BASKET FREE AREA
		A	K	L	LBS.	85.0 SQ. IN.
H-72	H-72-F	2	6	4¾	65	
H-73	H-73-F	3	7½	6	65	
H-74	H-74-F	4	9	7½	65	

Figure 9-12 A sediment or hair interceptor. Chart shows pipe size requirements for various installations. *(Courtesy Josam Mfg. Co.)*

discharged into drainage piping only after use of an approved neutralizing device or substance. This approval must come from the local plumbing code administrative authority.

9-11 What wastes require indirect waste piping?

Certain wastes require disposal through the installation of indirect waste piping. Included in this category are wastes from:

- Refrigerators
- Ice boxes

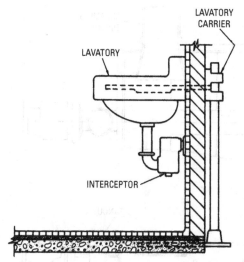

Figure 9-13   A typical lavatory sediment or hair interceptor.
(Courtesy Josam Mfg. Co.)

- Steam tables or other devices in which food and drink are stored or processed
- Devices such as stills, sterilizers, and water cooling equipment
- Dishwashing machines except those used in private living quarters
- Overflows, drains, or relief vents from water supply systems

9-12 What type of piping installations are required for indirect wastes such as those just listed?

Wastes such as those listed shall discharge into the structure drainage system through indirect piping and be provided with an air gap of at least twice the effective diameter of the drain served. Provision can be made by extending the indirect waste pipe to a laundry sink or similar fixture, or by providing an air gap in the drain connection on the inlet side of the trap servicing the device or fixture.

9-13 What are considered to be *hot* wastes?

Hot wastes are those with temperatures generally above 180°F. These wastes are not permitted in a building sewer until the temperature falls below the stated 180°F range.

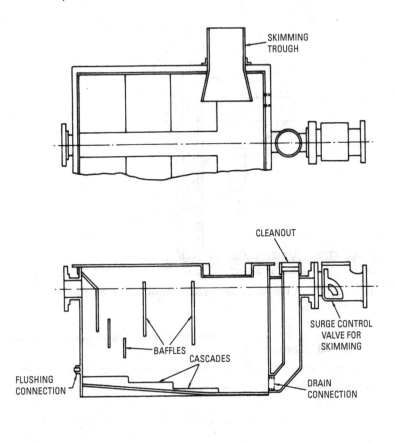

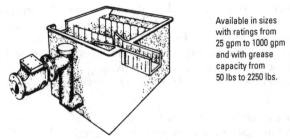

Available in sizes
with ratings from
25 gpm to 1000 gpm
and with grease
capacity from
50 lbs to 2250 lbs.

Figure 9-14 A typical packing-house grease interceptor with surge control valve and skimming trough. *(Courtesy Josam Mfg. Co.)*

9-14 Can waste water piping from swimming and wading pools, swimming pool scum gutters, or floor drains be directly installed into the drainage system?

Wastes from swimming pools and their drainage accessories shall be processed through indirect waste pipe with an approved air gap.

9-15 What is a sump pump?

A *sump pump* is an automatic water pump that is powered by an electric motor for removal of drainage, other than raw sewage, from a pit or low point by mechanical means (see Figure 9-15).

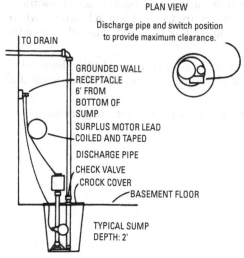

PLAN VIEW

Discharge pipe and switch position
to provide maximum clearance.

TO DRAIN

GROUNDED WALL
RECEPTACLE
6' FROM
BOTTOM OF
SUMP

SURPLUS MOTOR LEAD
COILED AND TAPED

DISCHARGE PIPE

CHECK VALVE

CROCK COVER

BASEMENT FLOOR

TYPICAL SUMP
DEPTH: 2'

Figure 9-15 A typical sump-pump installation. *(Courtesy Genova, Inc.)*

True or False Exercises

Place a T or F in the blank to the left of the question. (See Appendix C for answers.)

_____ **1.** It is required that a grease interceptor be installed on waste lines leading from sinks, drains, or other fixtures of restaurants, hotel kitchens and bars, clubs, factory restaurants or cafeterias, hospital kitchens, or any establishment from which excessive amounts of grease may be introduced into the sewer system.

_____ **2.** A grease interceptor is not required to be installed for an individual dwelling unit or for private living quarters.

_____ **3.** If an establishment has the capacity to serve 10 or more meals a day, a grease interceptor is considered necessary.

_____ **4.** The installation of a grease interceptor should be such as to protect it from freezing.

_____ **5.** Interceptors may not necessarily be readily accessible for cleaning.

_____ **6.** All codes require the periodic removal of accumulated grease to keep interceptors in an efficiently operating condition.

_____ **7.** If more than one fixture discharges into an interceptor, only one fixture shall be trapped and vented.

_____ **8.** Codes require that interceptors be installed in locations where motor vehicles are serviced and in locations where excessive sediment and/or hair may enter the drainage lines.

_____ **9.** Harmful wastes are wastes considered to be harmful to the piping of a plumbing system and include acids and alkalines.

_____ **10.** Hot wastes are those with temperatures generally above 180°F.

Chapter 10

Drainage, Vents, and Vent Piping

Drainage is an important part of waste disposal in areas of human habitation. Plumbers are not concerned, in this instance, with land surface drainage, but only with the disposal of sewage, wastes, and drainage from roof areas. The proper installation of pipe, fixtures, vents, and drainage systems is an important factor in maintaining the health and well-being of building occupants.

Generally speaking, the drainage system comprises all parts of the system (including vents) inside the building. Drainage installed outside the building is normally referred to as the *sewer system* and consists of either a sanitary-type system or a storm sewer that collects exterior surface water drainage.

Figure 10-1 illustrates a typical building sanitary drainage pipe connected to the main street sanitary sewer. On small installations, one main sanitary drain from a building is sufficient, but on large buildings or residential apartment complexes, more drains may be required because of the scattered locations of the numerous fixtures served.

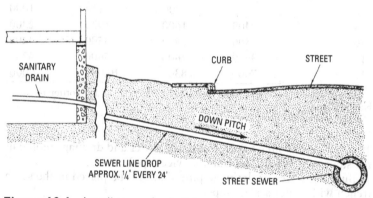

Figure 10-1 Installation of a building sanitary drain connected to a street sanitary sewer.

The following questions and answers are typical of those asked in the various plumbing licensing examinations.

10-1 What are the general rules concerning building drainage piping?

Horizontal drainage piping shall be run at a practical alignment and at a uniform slope. The minimum allowable size of any building drain, horizontal branch, vertical soil, or waste stack serving one or more water closets shall be 3 inches.

The building sewer and the main building drain are not recommended to be less than 4 inches in diameter. Branches of sewers and drains are sized in accordance with a table of allowances, or in accordance with good engineering practices. Table 10-1 illustrates typical fixture allowances in relationship to drainage pipe sizes.

Table 10-1 Typical Fixture Allowances in Relationship to Drainage Pipe Sizes*

| Pipe Diameter (In Inches) | Fall or Slope per Foot | | | |
	$1/_{16}$ Inch	$1/_8$ Inch	$1/_4$ Inch	$1/_2$ Inch
2			21	26
2$1/_2$			24	31
3		20†	27†	36†
4		180	216	250
5		390	480	575
6		700	840	1000
8	1400	1600	1920	2300
10	2500	2900	3500	4200
12	3900	4600	5600	6700
15	7000	8300	10,000	12,000

*Maximum number of fixture units that may be connected to any portion of building drains. (Chart is only typical.)
†Not more than two water closets.

10-2 Is it permissible to place building sewer and drainage piping in the same trench with the water service pipe?

Building sewer or drainage pipe shall *not* be placed in the same trench with water service pipe.

Sewer and drainage pipe shall *not* be less than 10 feet apart horizontally and shall be separated by undisturbed or compacted earth except when installed in utility tunnels. Special permission for installation in other than recommended procedures may be requested from the local code administrative governing authority.

10-3 How are direction changes made in drainage piping?

Direction changes in drainage piping shall be made by the use of 45° wyes, long-or-short sweep quarter bends, eighth or sixteenth bends, or by a combination of these or equivalent fittings. When threaded-type drainage fittings are used, they shall be of the recessed type.

10-4 May crosses and tees be used for drainage-system pipe fittings?

Crosses and tees shall not be used in the drainage system except when used for sinks or lavatories that are located back-to-back.

10-5 Are there any prohibited drainage pipe fittings and connections?

Tapping or drilling of drainage or vent piping and the use of fittings that have a hub in the direction opposite to flow are prohibited. Tee branches are prohibited for use as drainage fittings, as are the use of running threads, bands, and saddles. Drainage pipe must be protected against damage. Pipe crossing through or under walls shall be protected from damage and breakage.

All buried piping shall be supported throughout its entire length. When piping is laid in open trenches, backfill shall be properly compacted around the piping without damage to the piping. All fill shall be properly compacted with clean earth that shall *not* contain stones, boulders, cinder fill, or other materials that would damage or break the pipe or result in corrosive action.

Water, soil, and waste pipe installed outside of a building or in exterior walls shall be protected against freezing in localities where freezing may occur.

10-6 What is the purpose of a plumbing trap?

Traps are devices or fittings used to provide a liquid seal to prevent sewer gases from getting into a house or building. Figure 10-2 indicates how a trap works. Toilets have built-in traps because their intricate bowl passages are trap shaped.

10-7 What is a plumbing stack?

A *plumbing stack* is a vertical line of drain, waste, or vent pipe that vents above the roof. If the stack serves a toilet, it is called a *main stack* or *soil stack*. Figure 10-3 shows a typical drain, waste, and vent system.

10-8 How can building drains be discharged if they are below the gravity flow of the drainage system?

Below–gravity flow building drains shall be discharged into a gas-tight and vented sump pump so that the liquid can be mechanically lifted and discharged into the building gravity-flow drainage system.

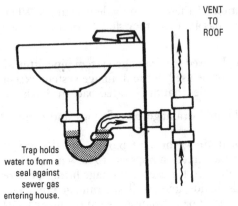

VENT
TO
ROOF

Trap holds
water to form a
seal against
sewer gas
entering house.

Figure 10-2 How a trap works. *(Courtesy Genova, Inc.)*

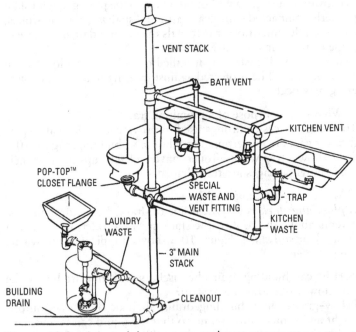

VENT STACK

BATH VENT

KITCHEN VENT

POP-TOP™
CLOSET FLANGE

SPECIAL
WASTE AND
VENT FITTING

TRAP

LAUNDRY
WASTE

KITCHEN
WASTE

3" MAIN
STACK

BUILDING
DRAIN

CLEANOUT

Figure 10-3 A typical drain, waste, and vent system. *(Courtesy Genova, Inc.)* .

10-9 What are some of the types of floor drains and what are the recommended minimum sizes for floor drains and traps installed below a basement floor?

There are a variety of floor-drain grates, some of which are illustrated in Figure 10-4.

Figures 10-4A and B illustrate floor grates recommended for installation where heating condensate lines deposit waste into a floor drain or sink in an exposed area. These floor drains or sinks are available with either $\frac{1}{2}$ or $\frac{3}{4}$ grates.

Figure 10-4C shows a depressed grate or drain designed for use when removal of a sediment bucket is restricted; thus, the depressed grate serves as a bucket to catch debris (generally installed under counters in food markets, restaurants, and bars).

Figure 10-4D shows an angle grate drain that is ideal for installation in restaurants where drainage is required under a counter or in a service area. The angle grate provides maximum drainage plus easy serviceability.

Figure 10-4E shows a center-hole grate drain that allows installation of indirect waste lines and prevents splashing while providing secondary drainage.

Figure 10-4F illustrates a funnel with a grate drain. This type of drain combination provides the feature of an indirect waste drain and floor sink. Funnels are available in two styles. The round funnel is for single-pipe discharge and the oval funnel is for multiple-pipe discharge.

Floor drains and floor drain traps installed underground or below a basement floor shall be a minimum of 2 inches in diameter. The drain inlet shall be accessible and in full view. Figure 10-5 shows a typical floor drain installation diagram.

10-10 Should a plumber installing drainage piping be concerned about the structural members of the building and about rodent-proofing?

Structural members of a building shall not be weakened by cutting, notching, and so on. Cuts and notches must be properly reinforced. Any section of the building that must be changed or replaced shall be left in a safe structural condition. Exterior pipe openings shall be rodent-proofed (rat-proofed) with well-fitting collars or metal securely fastened into place. Interior pipe openings through walls, ceilings, and floors shall be similarly protected.

10-11 What are the requirements for toilet-room ventilation?

Toilet rooms shall have ventilation by a window or exhaust ductwork. Where ducts are installed, exhaust shall be to the outer air or to an independent exhaust duct system.

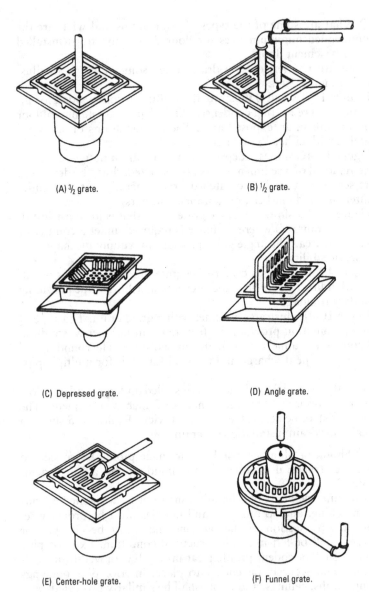

(A) ³/₂ grate.

(B) ¹/₂ grate.

(C) Depressed grate.

(D) Angle grate.

(E) Center-hole grate.

(F) Funnel grate.

Figure 10-4 Various types of floor drains. *(Courtesy Josam Mfg. Co.)*

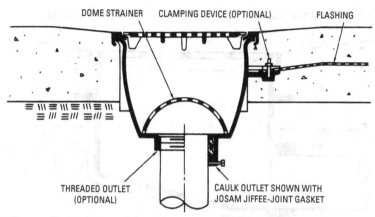

Figure 10-5   A typical floor drain installation diagram. *(Courtesy Josam Mfg. Co.)*

10-12 Are special drains required for swimming pool installations?
Under normal circumstances, special drains are required for swimming pools. Figure 10-6 shows a typical swimming pool scum-gutter drain. Figure 10-7 illustrates an installation diagram of a main outlet trench drain. Figure 10-8 shows an installation diagram of an interlocking vandal-proof grating in a trench drain.

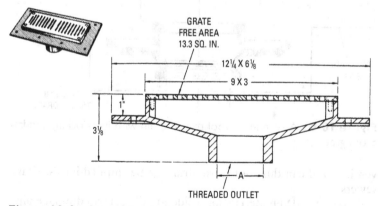

Figure 10-6   A typical swimming pool scum-gutter drain.

10-13 What conductor or leader pipe for roof drainage is used for installation on the interior of buildings? May these pipes be used for

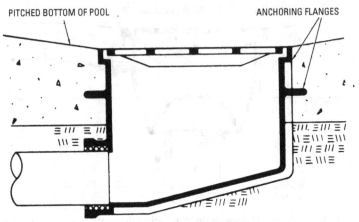

Figure 10-7 A swimming pool main outlet drain.

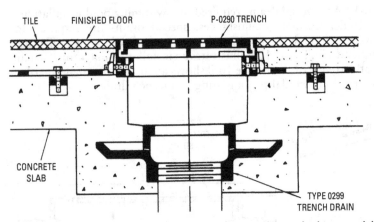

Figure 10-8 A swimming pool trench drain with interlocking vandal-proof grating.

venting, and can this roof storm drainage be emptied into sanitary sewers?

Recommended materials for inside-building storm drainage pipe installation include galvanized steel, galvanized wrought iron, cast iron, brass, lead, and copper pipe. Under *no* circumstances can this pipe be used for soil, waste, or vent pipe, nor shall any soil, waste, or vent pipe be used for a storm drainage conductor pipe. In addition, the passage of storm drainage into sanitary sewers is prohibited.

10-14 Are traps required for storm water drains if they are connected to sewers carrying only storm water, and are connections of building floor drains to storm sewer drains permitted?

Traps are *not* required for storm water drains if they are connected to sewers carrying only storm water. *Connection of building floor drains to storm sewer piping is prohibited.*

10-15 What procedures or calculations are generally used for sizing building storm-drainage conductor pipe?

Calculations are generally based on the total of roof area, the maximum amount of rainfall per hour in the area, and the roof slope. Table 10-2 illustrates a typical conductor-pipe sizing chart based upon a maximum rainfall of 4 inches per hour. Figure 10-9 illustrates a typical flat-roof surface conductor pipe roof drain; Figure 10-10 shows a typical flat-roof drain installation diagram.

Table 10-2 Typical Conductor-Pipe Sizing for Vertical Roof Area

Conductor Pipe Diameter in Inches	Projected Roof Area in Square Feet
2	700
$2\frac{1}{2}$	1250
3	2100
4	4500
5	8400

Note: Table based on a maximum rainfall of 4 inches per hour.

10-16 Do all buildings in which plumbing fixtures are installed require soil or waste stack vents?

All buildings having plumbing fixtures installed shall have a soil or waste stack (or stacks) extending full-size through the roof or as approved by the local plumbing-code administrator. Soil and waste stacks shall be as direct as possible, free of sharp bends or turns, and shall be flashed against entrance of rain and snow. No water closet or closets shall discharge into a stack of less than 3 inches in diameter.

10-17 What type of pipe materials are generally used for soil, waste, and vent piping?

Pipe and fittings materials used for soil, waste, and vent piping are of galvanized steel, cast iron, stainless steel, galvanized wrought iron, lead, brass, copper, or plastic, and must conform to the approval of the local code administrative authority.

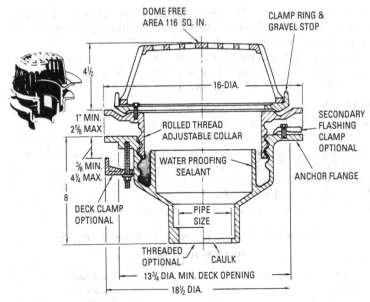

Figure 10-9 A typical flat-roof conductor-pipe drain.

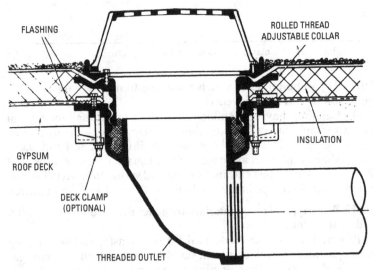

Figure 10-10 A flat-roof conductor-pipe drain installation diagram.

10-18 What type of fittings are used for plumbing vents?

Fittings shall conform to the type of pipe material used in the vent system. Some codes allow the use of galvanized malleable-iron and cast-iron steam pattern fittings with threaded-type pipe.

10-19 How are fixture trap seals protected?

Trap seals are protected from back pressure and siphonage by the installation of waste or soil stack vents, back vents, revents, or circuit or continuous vents.

10-20 What is the general procedure for installing ventilation stacks?

If a roof rafter or other obstruction requires turning the stack, an offset can be made (see Figure 10-11). When such an offset occurs in the drainage part of a stack, use a 45°L (¹⁄₈ bend); if it occurs in the vent portion of the stack, use a 90°L (¹⁄₄ bend). The topmost length of a stack (called the *increaser*) should extend at least 12 inches above the roof (see Figure 10-12). Roof flashing, shaped to lie flat on the roof slope and encircle the increaser, is installed to make the stack weathertight (see Figure 10-13).

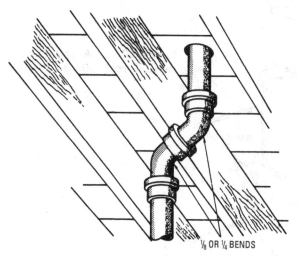

⅛ OR ¼ BENDS

Figure 10-11   A method used for offset in vent stack.

10-21 Is a main vent or vent stack required for two or more branch sections?

A main vent or vent stack shall be installed with a soil or waste stack if relief vents, back vents, or other branch vents are required in two or more branch sections.

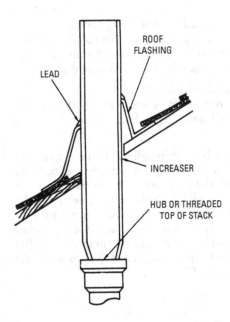

Figure 10-12 A vent increaser roof extension.

ROOF FLASHING

LEAD

INCREASER

HUB OR THREADED TOP OF STACK

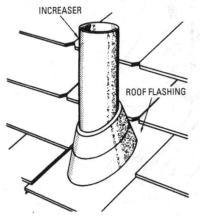

INCREASER

ROOF FLASHING

Figure 10-13 A typical vent increaser roof flashing.

10-22 Can frost or freezing result in poor or no venting in soil and waste stacks?

Frost or freezing can result in poor or closed venting. No soil or waste stack shall be installed or permitted outside a building unless protected against frost. Vent terminals shall not be screened in frost areas. If a possibility exists of closure by frost, vents shall be increased in size to sufficient diameter to prevent frost closures.

10-23 Are there any prohibited terminal stack or vent locations?

No vent pipe shall terminate under the overhang of a building. Roof terminals of stacks and vents, if located within at least 6 feet of any window, door, scuttle, or air shaft, shall be extended at least a minimum of 2 feet above such a location.

10-24 What types of fittings are used in branch vent systems?

The fittings used in branch systems are the same as the fittings used for water or steam, except that they are made with a shoulder so that their inside diameter is the same as the pipe (see Figure 10-14). The inside surface of the fittings comes flush with the inside surface of the pipe so that there is no projecting part to catch particles of matter and clog the passage.

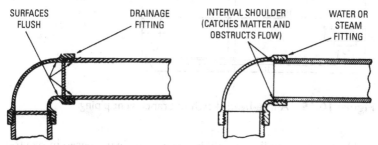

SURFACES FLUSH   DRAINAGE FITTING   INTERVAL SHOULDER (CATCHES MATTER AND OBSTRUCTS FLOW)   WATER OR STEAM FITTING

Figure 10-14   Typical drainage fittings.

10-25 Make a rough free-hand sketch of a residential vent pipe system and include pipe sizes.

Measurements are not required for a free-hand sketch, but approximate locations of the fixtures should be shown. The general arrangement of piping is illustrated simply by lines drawn in free-hand, as shown in Figure 10-15.

10-26 What is the proper procedure to follow in laying out branch vent lines?

Branch vent lines should pitch upward from the trap to the connection with the vent stack. The point at which a vent line enters the stack should be sufficiently higher than the trap that it serves so that there is no possibility of a backflow of water through the vent line into the vent stack. Each vent line should have a separate connection with the vent stack.

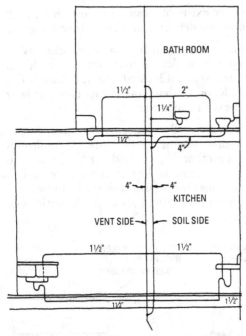

Figure 10-15 A free-hand sketch of branch vent piping.

The branch vent line should be progressively increased in size toward the stack in proportion to the number of vent pipes along the line, as shown in Figure 10-16. The following is the method of proportioning the branch vent. Starting at the toilet (*A*), the minimum-size vent pipe permissible is 2 inches. This 2-inch vent pipe will connect with the branch vent at (*B*), preferably by a long-turn elbow (to reduce the resistance of the air flow). The branch vent will be the same size pipe (2 inches) up to the junction at (*C*), where the lavatory pipe is connected. Here, the branch vent should be enlarged. The size of the vent between (*C*) and the stack is obtained as shown in Table 10-3.

There is no size of standard wrought pipe (the kind used in this instance) corresponding to this area:

Internal area $2\frac{1}{2}$-inch pipe = 4.788 in^2

Internal area 3-inch pipe = 7.393 in^2

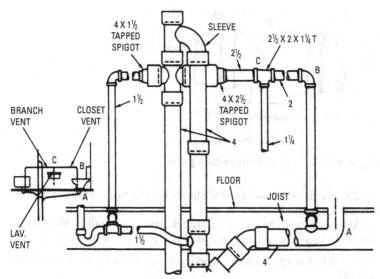

Figure 10-16 An illustration of a branch vent line installation.

Table 10-3 Calculation of Total Transverse Area

Internal transverse area	
2-inch closet vent pipe	3.355 in²
1½-inch lavatory vent pipe	2.036 in²
Total transverse area to be served	5.391 in²

In accordance with these calculations, the 2½-inch pipe is too small and the 3-inch pipe too large, so a choice between the two must be made. Since both traps seldom discharge at the same time, the smaller (or 2½-inch pipe) is selected. The junction at (C), where the branch vent is enlarged, is made with a reducing tee, as shown in Figure 10-17. It is not a good practice to run a branch vent the same size to the stack, regardless of the number of traps served.

10-27 May a vent stack be used as a support for a television antenna or other roof-placed extensions?

Using vent terminals as supports for television antennas, flag pole supports, or similar purposes, is prohibited.

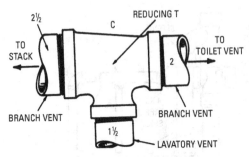

Figure 10-17 A tapering reducing tee.

10-28 Are there recommended slopes for the installation of vent or branch vent pipe?

All connections of vent and branch vent pipe shall be sloped and connected so as to drip by gravity back to the soil or waste pipe.

10-29 What is the recommended average distance of a trap from a vent?

General maximum recommended distances of any waste pipe from a fixture to a stack including the vertical drop from the trap seal to the outlet of the waste are as shown in Table 10-4.

Table 10-4   Maximum Recommended Distance of any Waste Pipe from a Fixture to a Stack

Pipe Diameter	Maximum Distances
1¼ inches	4 feet
1½ inches	5 feet
2 inches	5 feet
3 inches	6 feet
4 inches	8 feet

These are the average recommended distances. If an approved drum trap is used on a bath waste, an additional 3 feet of distance is allowed. If an approved centrifugal trap is used on a fixture other than a bathtub or water closet, an extension of 3 feet in distance is allowed over and above listed distances.

10-30 Is a common vent permissible for use with two fixtures?

A common vent may generally be used for two fixtures (excluding water closets) if they are on the same floor level. Connections must be made at different stack levels with the vertical drain being one pipe diameter larger than the size of the upper fixture drain and in accordance with local plumbing code requirements.

10-31 What are the recommended sizes of individual relief vents and of average individual vents?

The recommended diameter of an individual relief vent is not less than one-half the diameter of the waste branch or soil pipe to which the connection is to be made. Individual vents shall be not less than $1\frac{1}{4}$ inches in diameter or less than one-half the diameter of the drain connection.

10-32 May charts or tables be used to determine vent pipe sizes?

Charts or tables may be used to determine vent pipe sizes and lengths. These charts can be found in the various state plumbing codes and in approved technical books. Table 10-5 illustrates a typical vent pipe size and length determination chart. First-step procedures include a determination of the total number of fixtures to be installed that require venting and then the length of the venting. Normally 20 percent of the total vent pipe may be installed in a horizontal position.

Table 10-5   Typical Vent Size and Length Determination

Stack Size— Soil or Waste	Connected Fixture Units	$1\frac{1}{4}$	$1\frac{1}{2}$	2	$2\frac{1}{2}$	3	4
		In Feet, Maximum Length of Vent					
Inch Size							
$1\frac{1}{4}$	2	30					
$1\frac{1}{2}$	8	50	150				
$1\frac{1}{2}$	10	30	100				
2	12	30	75	200			
2	20	26	50	150			
$2\frac{1}{2}$	42		30	100	300		
3	10		30	100	200	600	
3	30			60	200	500	
3	60			50	80	400	
4	100			35	100	260	1000
4	200			30	90	250	900

In Inches, Required Diameter of Vent

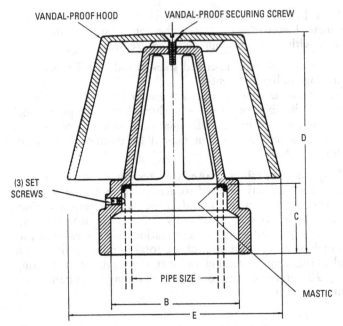

VANDAL-PROOF HOOD

VANDAL-PROOF SECURING SCREW

(3) SET SCREWS

D

C

PIPE SIZE

B

E

MASTIC

Figure 10-18 A cast-iron vandal-proof hooded vent.

10-33 Are vandal-proof hooded vent caps ever installed on vent pipes?

In certain locations where roofs are accessible to unauthorized persons, vandal-proof hooded vent caps are installed to ensure proper venting and to prevent the introduction of unauthorized objects into the stack. Figure 10-18 shows a typical cast-iron vandal-proof hooded vent cap. Figure 10-19 illustrates an installation diagram of a cast-iron vandal-proof hooded vent cap.

10-34 What is a vent system?

A *vent system* is a pipe or series of pipes installed to provide air flow to or from a drainage system or to provide circulated air within the system to protect trap seals from back pressure and siphonage (see Figure 10-20).

10-35 What is a hand-operated closet auger?

A *hand-operated closet auger* is a tool (device) used to open clogged drains and toilets (see Figures 10-21 and 10-22).

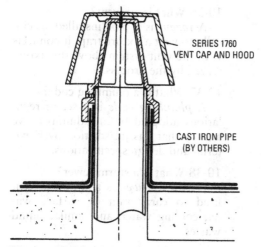

SERIES 1760
VENT CAP AND HOOD

CAST IRON PIPE
(BY OTHERS)

Figure 10-19 An installation diagram of a cast-iron vandal-proof hooded vent cap. *(Courtesy Josam Mfg. Co.)*

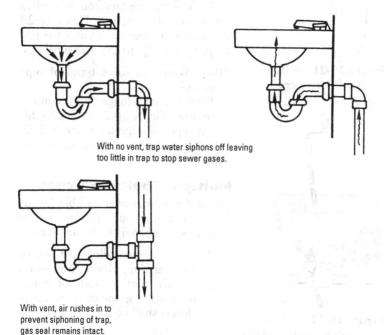

With no vent, trap water siphons off leaving
too little in trap to stop sewer gases.

With vent, air rushes in to
prevent siphoning of trap,
gas seal remains intact.

Figure 10-20 How a vent system works. *(Courtesy Genova, Inc.)*

Figure 10-21 A hand-operated closet auger.

(Courtesy Electric Eel Mfg. Co., Inc.)

Figure 10-22 A hand-operated closet auger in use.

(Courtesy Electric Eel Mfg. Co., Inc.)

10-36 What is a revent?

A *revent* is a pipe installed specifically to vent a fixture trap. It connects with the vent system above the fixture served by the process.

10-37 What is a plumbing code?

A *plumbing code* is a set of regulations adopted by an administrative agency that has jurisdiction over material and design specifications.

10-38 What is a storm sewer?

A *storm sewer* is a drainage vent used to carry rain or surface water, cooling water, and similar liquid wastes.

10-39 What is an electric sanitary plumbing rod or auger?

An *electric sanitary rod* is used to open drainage stoppages. Figure 10-23 shows one of these devices used to open stoppages in $1\frac{1}{4}$- to $2\frac{1}{2}$-inch pipelines.

10-40 What are some types of pipe cleaners?

Pipes become clogged for a number of reasons. Figure 10-24 shows 16 different types of augers, cutters, a chain knocker, and a brush used on various types of clogs.

Multiple-Choice Exercises

Select the right answer and blacken *a*, *b*, *c*, or *d* on your practice answer sheet. (See Appendix C for answers.)

1. The minimum allowable size of any building drain, horizontal branch, vertical soil, or waste stack serving one or more water closets shall be _____.

   a. 3 inches

   b. 4 inches

TOILET
GUIDE TUBE
AND HOLDER

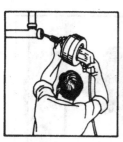

RODDING OVERHEAD
PIPELINES

OPENING KITCHEN
SINKS, FLOOR
DRAINS, SHOWER
STALLS, ETC.

REMOVING TOILET
OBSTRUCTIONS

Figure 10-23   An electric plumbing auger used to open clogged waste piping and to remove toilet obstructions. *(Courtesy Electric Eel Mfg. Co., Inc.)*

    c. 6 inches

    d. 1³/₄ inches

2. Building sewer or drainage pipe shall _____ be placed in the same trench with water service pipe.

    a. or can

    b. or will

    c. not

    d. may

3. Direction changes in drainage piping shall be made by using _____ wyes, long- or short-sweep quarter bends, eighth or sixteenths ends, or by a combination of these or equivalent fittings.

    a. 60°

    b. 45°

    c. 90°

    d. elbow

4. Which one of the following is not a prohibited drainage pipe fitting or connection?

    a. Tapping or drilling of drainage or vent piping.

    b. Fittings having a hub in the direction opposite to flow.

DROP HEAD AUGER For cleaning back-to-back mounted fixtures e.g., sinks where cable needs to be led into down pipe.

STRAIGHT AUGER For use in exploring and breaking up stoppages or returning sample to surface to determine correct tool.

FUNNEL AUGER For use as second tool in line. Breaks up remains of stoppage left by straight auger.

HOOK AUGER For heavy and dense root stoppages in pipes that require hooking and breaking up.

RETRIEVING AUGER For searching for cable that is broken or lost in line.

GREASE CUTTER For lines that have become badly greased with detergents and have to be opened.

FOUR BLADE SAW TOOTH CUTTER For following up after augers have been used and to open up floor drains.

SPADE CUTTER For blockages caused by hardened, glazed material such as chemical deposits.

Figure 10-24 Pipe cleaners. (Courtesy *Ridgid*)

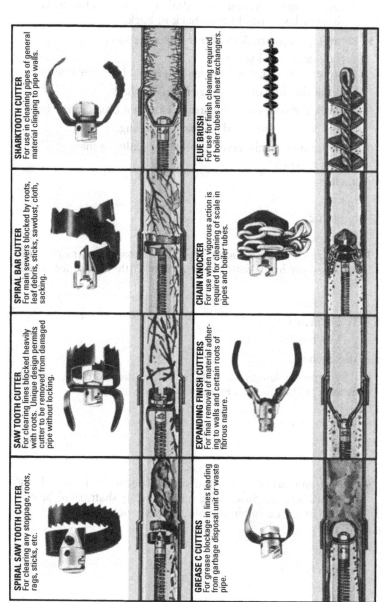

SPIRAL SAW TOOTH CUTTER
For clearing any stoppage, roots, rags, sticks, etc.

SAW TOOTH CUTTER
For clearing lines blocked heavily with roots. Unique design permits cutter to be removed from damaged pipe without locking.

SPIRAL BAR CUTTER
For main sewers blocked by roots, leaf debris, sticks, sawdust, cloth, sacking.

SHARKTOOTH CUTTER
For use in cleaning pipes of general material clinging to pipe walls.

GREASE C CUTTERS
For grease blockage in lines leading from garbage disposal unit or waste pipe.

EXPANDING FINISH CUTTERS
For final removal of material adhering to walls and certain roots of fibrous nature.

CHAIN KNOCKER
For use when vigorous action is required for cleaning of scale in pipes and boiler tubes.

FLUE BRUSH
For use for finish cleaning required of boiler tubes and heat exchangers.

Figure 10-24 *(Continued)*

c. Tee branches for use as drainage fittings.

d. Crosses and tees used in the drainage system when used for sinks or lavatories located back to back.

5. Plumbing traps are devices or fittings used to _____.

a. provide a liquid seal to prevent sewer gases from getting into a house or building.

b. check the flow of hot water in an overflow.

c. check the flow of cold water in case of a break in the line.

d. prevent overflows and excessive draining.

6. If a plumbing stack serves a toilet, it is called a main stack or _____ stack.

a. clay

b. cast

c. plastic

d. soil

7. The size of the conductor pipe, for a maximum rainfall of 4 inches per hour, with a roof area of 2100 square feet shall be _____.

a. 2 inches

b. 2½ inches

c. 3 inches

d. 5 inches

8. The topmost length of a stack, called the increaser, should extend at least __ inches above the roof.

a. 9

b. 12

c. 36

d. 48

9. Roof terminals of stacks and vents, if located within at least 6 feet of any window, door, scuttle, or air shaft, shall be extended at least a minimum of __ above such a location.

a. 3 feet

b. 4 feet

c. 5 feet

d. 2 feet

10. The recommended average distance of a trap from a vent is
____.
   a. 4 feet for a $1^{1}/_{4}$-inch pipe
   b. 6 feet for a $1^{1}/_{2}$-inch pipe
   c. 8 feet for a 2-inch pipe
   d. 10 feet for a 4-inch pipe

Chapter 11

Water Supply and Distribution

Plumbing codes maintain strict requirements for water supply distribution systems. Codes state that the water supply distribution system for drinking, culinary, and domestic purposes shall be distributed through a piping system entirely independent of any piping system supplying water that is not approved for human consumption.

No plumbing devices or connections shall be installed in any water supply system that provides a cross-connection between a water supply system for drinking and domestic purposes, and a drainage system or a soil or waste pipe that would allow the backflow of waste or sewage into the distribution system. Water supply connections to hospital sterilizers, bedpan sterilizers, house tanks, plumbing fixtures, and swimming pools shall be so made as to prevent the return of any of the liquid or waste matter either by gravity or siphonage to the water supply distribution system.

The following are typical questions asked in licensing examinations.

11-1 May nonpotable water be used for any plumbing fixture?

Nonpotable water may be used for flushing urinals and water closets. Such water shall not be accessible for drinking or domestic purposes, and there shall be no physical connection between potable and nonpotable supply systems. Nonpotable water supply piping shall be identified by color-coding in accordance with an approved scheme for identification of piping systems. The preferred code is U.S.A. 253.1.

11-2 What are the usual supply sources for domestic-use water?

Water for domestic use is usually obtained from the following:

- Wells
- Springs
- City mains

Wells are dug, bored, driven, or drilled. A *dug well* is constructed by excavating a shaft and installing a casing. Dug wells are generally of shallow construction. Bored wells are built by boring a hole with a hand- or machine-driven auger. These augers range in diameter from a few inches to as much as 4 feet or more where machine-driven augers are used. *Driven wells* are made by driving a pipe equipped with a well point by means of a maul or pile driver. Pipes used are usually 2 inches or less in diameter. *Drilled wells* are made by making a hole with a drilling machine and installing casing and a screen.

Supply sources in cities are from a street pressure system (the street main) into which water is furnished under pressure by the city supply source. A service pipe is run from the main to the building and connects to the supply pipe; suitable branches are run to the various fixtures (see Figure 11-1).

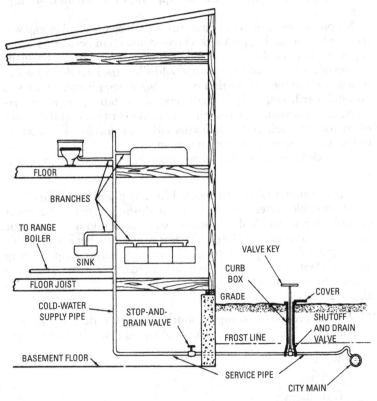

Figure 11-1 A typical cold-water supply pipe connected to a city main.

11-3 Are there requirements for protecting potable water supply tanks?

Potable water supply tanks shall be covered to protect them from entrance of rodents, insects, and other foreign materials. In addition, properly sized screen overflow pipes must be installed.

11-4 What are the requirements for materials used inside water supply tanks?

Materials used on the inside of domestic water supply tanks shall not affect the taste or *potability* of the water. During repair or cleaning, the tanks shall be disconnected from the system so as to prevent foreign materials or fluids from entering the water supply distribution system. Paints or coatings applied to the inside of potable water supply tanks shall be only those approved by the local code administrative authority.

11-5 What type of piping material is used in water distribution supply systems?

Materials used in water distribution systems include copper, brass, plastic, lead, block tin, tin-lined tubing, galvanized wrought iron or steel, stainless steel, and asbestos cement, all with approved fittings. Piping materials that were used in other than potable water supply systems are not permitted to be installed in potable water supply distribution systems. No materials that could produce toxic conditions in the piping installations are recommended for use.

11-6 Are there special considerations for water service lines to buildings?

In sizing water service lines to buildings, factors such as requirements for flushometers or other devices that need a high rate of water flow for operation must be considered. Service pipe shall be of adequate size to provide building water requirements during peak demands, and in no instance shall the service line be less than $3/4$ inch in nominal diameter.

11-7 Are valve controls required for the water distribution system within the building other than that provided for the main service entrance?

It is recommended that valve controls within a building be provided for each hot-water tank, water closet, urinal, and lawn sprinkler. These valves are not always required in single-family dwellings. However, in public buildings or multiple-dwelling units, a valve is recommended for control at the base of each riser, and for each dwelling unit or public toilet room, unless served by an independent riser. Control valves are also recommended for installation of fixtures isolated from a group.

On the exterior of the building, a main shutoff valve shall be installed near the street curb and gutter line.

11-8 What are the requirements, or remedy, for the suppression of noise or water pressure hazards in water distribution systems?

All water supply and distribution systems shall be provided with air chambers or other approved mechanical devices to suppress water-hammer line noises and to prevent pressure hazards to the

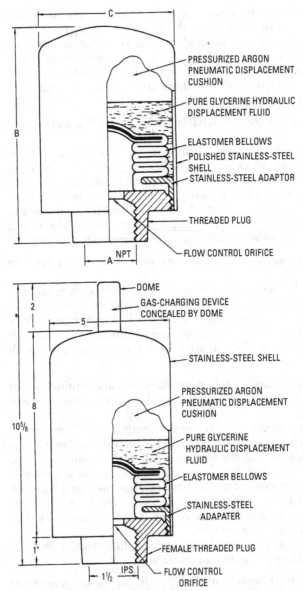

Figure 11-2 Hot-water and cold-water supply line shock absorbers.
(Courtesy Josam Mfg. Co.)

piping system. Mechanical water suppressors shall be accessible for inspection and repair. Figure 11-2 illustrates typical shock absorber devices; Table 11-1 shows the table of dimensions for these units. Figure 11-3 illustrates a typical location for these units in hot-water and cold-water supply branches.

Table 11-1 Dimensions for Hot-Water and Cold-Water Supply Line Shock Absorbers

Sizing Chart	Dimensions			Approximate Weight (lbs)
Fixture Units	A	B	C	
1–11	³/₄ inch	4¹/₄ inches	3¹/₄ inches	2 lbs, 12 oz
12–32	1 inch	5¹/₄ inches	3¹/₄ inches	3 lbs, 4 oz
33–60	1 inch	6 inches	3¹/₄ inches	3 lbs, 6 oz
61–113	1 inch	6³/₄ inches	3¹/₄ inches	3 lbs, 8 oz
114–154	1 inch	6³/₄ inches	5 inches	7 lbs, 5 oz
155–330	1 inch	7³/₄ inches	5 inches	8 lbs, 2 oz

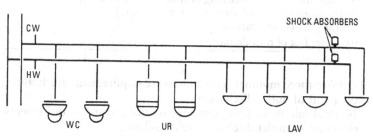

Figure 11-3 Typical location of shock absorbers in water line.

11-9 What is meant by the term fixture unit, and how was this term derived?

The unit system has been formulated from tests conducted by the Subcommittee on Plumbing and Building Codes under the Department of Commerce. Standard plumbing fixtures were installed and individually tested, and the amount of liquid waste that would be discharged through their outlet orifices in a given interval was carefully measured.

During the test, it was found that a washbasin (one of the smaller plumbing fixtures) would discharge waste in the amount of approximately 7¹/₂ gallons of water per minute. Since 1 cubic foot of water contains 7.4805 gallons, and because this volume was close to a cubic foot of water, the committee established this figure as the basis of the unit system and termed the discharge of the washbasin as one *fixture unit*. Therefore, one fixture unit is approximately 7½ gallons of water. Table 11-2 is a typical chart of fixture unit values.

Table 11-2   Chart of Typical Fixture Unit Values

Fixture	Units
Lavatory or wash basin	1
Kitchen sink	1½
Bathtub	2
Laundry tub	2
Combination fixture	3
Urinal	3
Shower bath	3
Floor drain	2
Slop sink	4
Water closet	6
One bathroom group (consisting of water closet, lavatory, bathtub, and overhead shower, or water closet, lavatory, and shower compartment)	8
180 square feet of roof drained	1

11-10 Are there minimal service pressure requirements for fixtures at the point of outlet discharge?

Normal minimum fairly constant service pressure requirements at the point of outlet discharge are as follows:

- Eight pounds per square inch minimum for all fixtures except direct-flush valves and for special equipment requiring higher pressures.
- Flush valve pressures shall not be less than 15 pounds per square inch at the point of discharge.
- Special equipment pressure requirements shall be in accordance with design needs.

NOTE

In all instances of minimum pressure requirements, consideration and allowance shall be made for pressure drop caused by friction loss in the system during high-demand cycles.

11-11 What is the effect of pipe friction loss in a water distribution system?

The friction loss in piping is an important factor and must be taken into account when evaluating a water distribution system. The friction loss shown in Table 11-3 is based on a section of 15-year-old pipe. With reference to this table, the friction loss through any one of the pipe sizes shown for any flow of water can be determined. For example, a glance at the friction loss table indicates that a discharge rate of 5 gallons per minute (gpm) through 100 feet of 1-inch iron pipe results in a friction loss of 3.25 feet. The same gpm through 100 feet of ³/4-inch pipe will result in a friction loss of 10.5 feet. From the foregoing, it will be noted that pipe friction must be taken into consideration when pipe is selected for the suction line on a shallow-well pump, or for the discharge pipe from the pressure tank to the point of delivery.

11-12 What are recommended nominal sizes for hot-water and cold-water fixture supply pipe?

Generally the diameter of any riser or branch serving more than one plumbing appliance or fixture shall be not less than ³/4 inch for iron or brass pipe. Branches to single fixtures shall not be less than ¹/2 inch in diameter. Three-eighths-inch pipe may be used to supply lavatories, water closet tanks, and drinking fountains if the supply pipe does not exceed 5 feet in length.

Table 11-4 illustrates a typical fixture pipe-sizing chart. Figure 11-4 shows an example of a domestic hot- and cold-water pipe-sizing sketch.

11-13 What are the pressure or temperature relief valve requirements for heating or hot-water vessels or units?

All domestic hot-water boilers or hot-water storage tanks of the closed type shall be provided with approved safety devices that meet the current requirements of a recognized testing organization, laboratory, or administrative authority. Typical safety devices include the following:

- A combination pressure and temperature relief valve
- An individual pressure relief valve and energy cutoff device
- Separate pressure relief and temperature relief valves

NOTE

Relief valves shall be installed at a distance of *not more than* 18 inches from a hot-water tank or boiler, and *no* shutoff valve of any type shall be placed between the relief valve and a hot-water boiler.

Table 11-3 Typical Water Supply Pipe Friction Loss

Flow (gpm)	Size of Pipe											
	1/2 Inch		3/4 Inch		1 Inch		1 1/4 Inches		1 1/2 Inches		2 Inches	
	FT	LBS	FT	LBS	FT	LBS	FT	LBS	FT	LBS	FT	LBS
2	7.4	3.2	1.9	0.82								
3	15.8	6.85	4.1	1.78	1.26	0.55						
4	27.0	11.7	7.0	3.04	2.14	0.93	0.57	0.25	0.26	0.11		
5	41.0	17.8	10.5	4.56	3.25	1.41	0.84	0.36	0.40	0.17		
6			14.7	6.36	4.55	1.97	1.20	0.52	0.56	0.24	0.20	0.086
8			25.0	10.8	7.8	3.38	2.03	0.88	0.95	0.41	0.33	0.143
10			38.0	16.4	11.7	5.07	3.05	1.32	1.43	0.62	0.50	0.216
12					16.4	7.10	4.3	1.86	2.01	0.87	0.70	0.303
14					22.0	9.52	5.7	2.46	2.68	1.16	0.94	0.406
16					28.0	12.10	7.3	3.16	3.41	1.47	1.20	0.520
18							9.1	3.94	4.24	1.83	1.49	0.645

Table 11-4 Fixture Pipe Sizing Chart

Fixture Type	Pipe Size	Fixture Type	Pipe Size
Bathtub	$1/2$ inch	Shower (single head)	$1/2$ inch
Drinking fountain	$3/8$ inch	Sink (flushing rim)	$3/4$ inch
Dishwasher (domestic)	$1/2$ inch	Urinal (Flushtank)	$1/2$ inch
Kitchen sink (residential)	$1/2$ inch	Urinal (direct-flush valve)	$3/4$ inch
Kitchen sink (commercial)	$3/4$ inch	Water closet (tank type)	$3/8$ inch
Lavatory	$3/8$ inch	Water closet (flush-valve type)	1 inch
Laundry tray (1, 2, or 3 compartment)	$1/2$ inch	Hose bibbs	$1/2$ inch

11-14 What are the types of water supply pumps and what are their typical characteristics?

Automatic and semiautomatic pump installations for water supply purposes generally employ four types of pumps, all conforming to the same general principles:

- The reciprocating, or plunger
- The rotary
- The centrifugal
- The jet or ejector (which receives its name from the introduction of a jet system attached to the centrifugal or reciprocating type of pump)

A brief tabulation of the characteristics of the various types of pumps is given in Table 11-5. For convenience, the pumps are classified according to speed, suction lift, and practical pressure head.

Careful consideration must be given to the selection of pumps for water supply purposes, because each pump application will differ (not only in capacity requirements, but also in the pressure against which the pump will have to operate). For example, if a $1/2$-inch hose with nozzle is to be used for sprinkling, water will be consumed at the rate of 200 gallons per hour. To permit use of water for other purposes at the same time, it is necessary to have a pump capacity in excess of 200 gallons per hour. Where $1/2$-inch hose with nozzle is to be used, it is necessary to have a pump with a capacity of at least 220 gallons per hour. In determining the desired pump capacity (even for ordinary requirements), it is advisable to select a size large

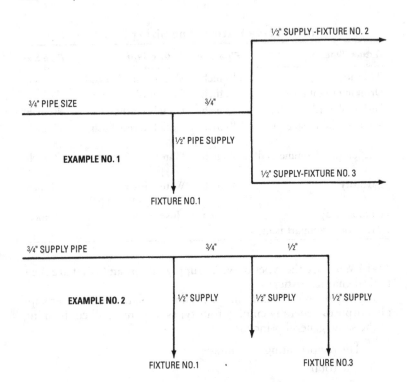

NOTE: Minimum pipe size to and from the water heater and to each fitting
where two or more ½" lines branch from the ¾" main shall be ¾".
½" ID rigid pipe or ½" OD flexible tubing shall be
used for fixture connections.

Figure 11-4 Examples of hot- and cold-water pipe sizing.

enough so that the pump will not run more than a few hours per
day.

Reciprocating pumps will deliver water in quantities proportional
to the number of strokes and the length and size of the cylinder. They
are adapted to a wide range of speeds and to practically any depth
of well. Since reciprocating pumps are positive in operation, they
should be fitted with automatic relief valves to prevent rupture of
pipes or other damage should power be applied against abnormal
pressure.

Centrifugal pumps are somewhat critical as far as speed is concerned and should be used only where power can be applied at a
reasonably constant speed. Vertical-type centrifugal pumps are used

Table 11-5 Pump Characteristics

Type of Pump	Speed	Practical Suction Lift	Pressure Head	Delivery Characteristics
Reciprocating				
Shallow well (low pressure)	Slow	22 to 25 ft	40 to 43 lbs	Pulsating (air chamber evens pulsations)
(medium pressure)	250 to 550 strokes per min		Up to 100 lbs	Pulsating (air chamber evens pulsations)
(high pressure)			Up to 350 lbs	
Deep well	Slow; 30 to 52 strokes per min	Available for lifts up to 875 ft; suction lift 22 ft below cylinder	Normal 40 lbs	Pulsating (air chamber evens pulsations)
Rotary Pump				
Shallow well	400 to 1725 rpm	22 ft	About 100 lbs	Positive (slightly pulsating)
Ejector Pump				
Shallow well and limited deep wells	Used with centrifugal-turbine or shallow-well reciprocating pump	Max. around 120 ft; practical at lifts of 80 ft or less	40 lbs (normal); available at up to 70 lbs pressure head	Continuous nonpulsating, high capacity with low-pressure head
Centrifugal				
Shallow well (single stage)	High; 1750 and 3600 rpm	15 ft maximum	40 lbs (normal); 70 lbs (maximum)	Continuous nonpulsating, high capacity with low-pressure head
Turbine Type				
Single impeller	High; 1750 rpm	28 ft maximum at sea level	40 lbs (normal); available up to 100 lbs pressure head	Continuous nonpulsating, high capacity with low-pressure head

in deep wells. They are usually driven through shafting by vertical motors mounted at the top of the well.

Turbine pumps, as used in domestic water systems, are self-priming. Their smooth operation makes them suitable for applications where noise and vibrations must be kept to a minimum.

Ejector pumps operate quietly, and neither the deep-well nor shallow-well type need be mounted over the well.

11-15 What are installation and capacity requirements for private water supply systems?

Private water supply systems shall be free of contamination and installed in a manner that conforms to the local administrative authority. Their accessories serving residences or public places shall be of a size (and possess a capacity with sufficient pressure and volume) to provide and maintain the plumbing in a safe and sanitary operating condition at all times.

11-16 What types of water hydrants are normally installed in parks and in outlying and rural farm areas?

Wall, box, and post water hydrants are installed in parks, as well as in outlying and rural farm areas, with an emphasis on nonfreeze types in cold weather locations. Figure 11-5 illustrates a typical nonfreeze wall water hydrant. Figures 11-6, 11-7, 11-8, and 11-9 show typical installation illustrations.

11-17 Can a simplified new top be placed on a valve box after repaving if the valve box is too low?

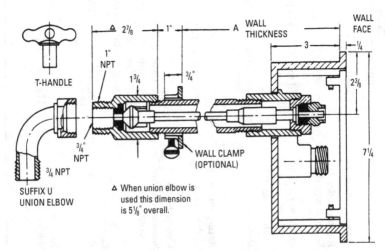

Figure 11-5 Typical nonfreeze wall water hydrant.

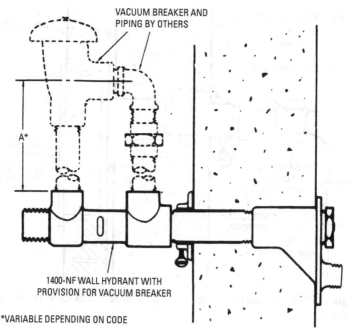

Figure 11-6   A nonfreeze hydrant for installation with a vacuum breaker.

A fitting (commercially named *Rite-Hite*) can be used by placing the fitting in the top of the valve box, pouring stiff cement mortar around it, and pushing it down until the fitting is flush with the pavement surface (see Figure 11-10 and Figure 11-11).

11-18 What is an air gap in the water supply system?

An *air gap* is the distance between the faucet outlet and the flood rim of the basin into which it discharges. An air gap prevents contamination of the water supply by back-siphonage.

11-19 Define potable water.

Potable water is water that is free from harmful impurities and is in conformity with the requirements of the U.S. Public Health Service Water Standards, or with the standards established by the local public health authorities.

11-20 What is a vacuum breaker in the water supply system?

A *vacuum breaker* in a water supply system is a device used to let in air and prevent back-siphonage.

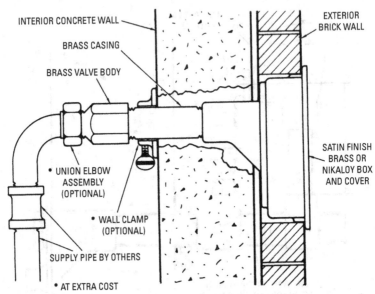

INTERIOR CONCRETE WALL

BRASS CASING

BRASS VALVE BODY

EXTERIOR
BRICK WALL

SATIN FINISH
BRASS OR
NIKALOY BOX
AND COVER

• UNION ELBOW
ASSEMBLY
(OPTIONAL)

• WALL CLAMP
(OPTIONAL)

SUPPLY PIPE BY OTHERS

• AT EXTRA COST

Figure 11-7 Installation diagram of a nonfreeze box-type water hydrant.

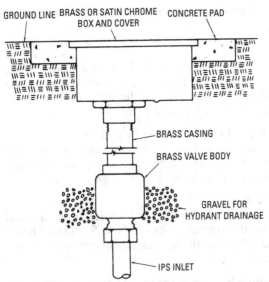

GROUND LINE BRASS OR SATIN CHROME   CONCRETE PAD
              BOX AND COVER

BRASS CASING

BRASS VALVE BODY

GRAVEL FOR
HYDRANT DRAINAGE

IPS INLET

Figure 11-8 Installation diagram of a nonfreeze box-type yard hydrant.

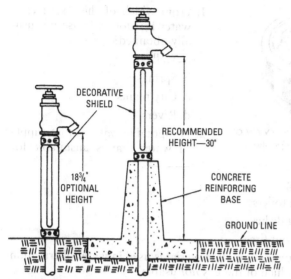

Figure 11-9 A nonfreeze post-type hydrant.

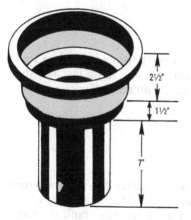

Figure 11-10 Valve box riser used to bring valve box to new pavement height after repaving.
(Courtesy General Engineering Co.)

11-21 What is a water main?
A *water main* is a water supply pipe for public use.

Multiple-Choice Exercises

Select the right answer and blacken *a*, *b*, *c*, or *d* on your practice answer sheet. (See Appendix C for answers.)

Figure 11-11 Top view of a valve box cover in place.

1. From which of these sources is water for domestic use *not* usually obtained?
   a. Wells
   b. Springs
   c. City mains
   d. Rivers

2. Nonpotable water is that supply of water that is unsuitable for _____.
   a. drinking
   b. flushing toilets
   c. washing dishes
   d. putting out fires

3. A copper type of piping material is used in water distribution systems. Which of these is *not* usable?
   a. Those that contain toxic substances
   b. Lead
   c. Plastic
   d. Stainless steel

4. Service pipe shall be of adequate size to provide building water requirements during peak demands, and in no instance shall the service line be less than _____ nominal diameter.
   a. 1 inch
   b. 1/4 inch
   c. 3/4 inch
   d. 2 inches

5. Control valves are required for the water distribution system within the building other than provided for in the main service entrance. They include valves for the hot-water tank, the water closet, the _____, and the lawn sprinkler.
   a. hot-water tank
   b. urinal
   c. shower
   d. laundry sink

6. Friction loss in a pipe accounts for slowing the rate of flow and can do the following to the amount of water delivered by a pump.

a. Decrease

b. Increase

c. Keep the rate the same

d. None of these

7. What size pipe should be used to connect to a shower?

a. $1/2$ inch

b. $1/4$ inch

c. $3/4$ inch

d. 1 inch

8. Automatic and semiautomatic pump installations for water supply purposes generally employ _____ types of pumps, all conforming to the same general principles.

a. three

b. two

c. one

d. four

9. If you are using a $1/2$-inch pipe to furnish water to a sprinkler system, the rate of flow of the water would be 200 gallons per hour. However, if a $1/2$-inch nozzle is used, it is necessary to have a pump with a capacity of at least _____ gallons per hour to provide the required 200 gallons per hour.

a. 300

b. 400

c. 220

d. 330

10. A vacuum breaker in a water supply is a device used to _____.

a. create a constant flow

b. let in air and prevent back-siphonage

c. prevent line breaks

d. eliminate uneven water flow

Chapter 12

Private Sewage Treatment Facilities and Residential Trailer Plumbing Standards

The average rural sewage disposal system consists of a septic tank complete with its tile disposal field. The word *septic* refers to any substance that produces or promotes the decomposition of animal or vegetable matter. Bacterial action is the basis of all septic tank systems. Waste from the bathrooms, kitchen, and laundry enters the septic tank. Through natural chemical processes, some of the mass is converted to gas, while the suspended solids settle to the bottom of the tank in the form of sludge.

The remaining liquid flows out of the tank through a series of conduits (the *tile field*). By distribution through the field, the liquids are eventually absorbed by the soil. Soil conditions, the building size, and the number of occupants are among the determining factors for selecting the tank and disposal field size. If the earth is composed of sand and gravel, and is consequently very absorbent, shorter tile fields will be satisfactory. However, if there is clay present to impede absorption, the disposal field must be larger to accommodate the liquid to be absorbed.

A septic tank is designed as follows:

- To provide a storage place where the motion of the liquid is arrested so as to give sufficient opportunity for the bacteria to reduce nearly all the solids to a liquid form.

- To provide a breeding place to increase the number of bacteria to accelerate the decomposition of the solids. The bacteria act strongly on vegetable and animal solids but cannot act on metal or mineral substances. This portion or sludge settles to the bottom of the tank and from time to time should be removed.

Built-on-site tanks are generally constructed as shown in Figure 12-1. After determination of the site, a hole is dug to such a depth that the tank will line up with the house sewer. The length and width of the excavation should be made at least 1 foot greater than the inside tank dimensions to provide space for the 6-inch concrete walls. Preconstructed septic tanks are usually installed since they are

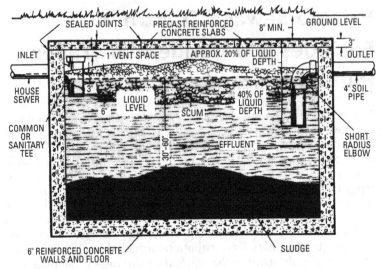

Figure 12-1 Sectional view of a typical septic tank.

readily available in most localities. Septic tank tile disposal fields are required for private sewage treatment facilities. Figure 12-2 illustrates a typical sewage septic tank disposal field layout. Figure 12-3 shows a typical trench layout for different soil conditions.

The following questions and answers are typical of those asked in licensing examinations in reference to private sewage treatment and residential trailer plumbing standards.

12-1 A four-bedroom home requires a 1200-gallon septic tank. If the floor of the tank is 4¹/₂ by 1¹/₂ feet, calculate the liquid depth and total depth of the tank.

Since the cubic content of the liquid equals approximately 1200/ 7.5, or 160 cubic feet, the depth of the contents will be 160/4.5 × 7.5, or 4.74 feet, which is approximately 4 feet 9 inches. From these calculations it can be determined that the inside dimensions of the tank will be as follows:

- *Inside width* —4 feet 6 inches
- *Inside length*—7 feet 6 inches
- *Liquid depth* —4 feet 9 inches
- *Total depth*—5 feet 9 inches

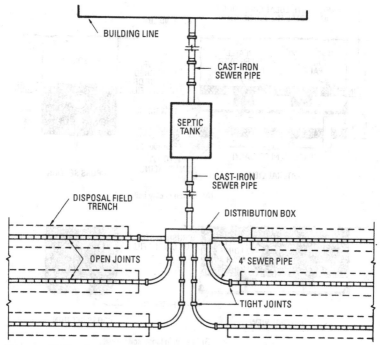

Figure 12-2   Sewage septic tank disposal field layout.

12-2 Can private sewage treatment facilities be installed if public sewage treatment facilities are available? What happens if public sewage treatment facilities become available after private treatment facilities are installed?

Private sewage treatment facilities may be installed only when no public sewerage system facilities are available and none will become available within a reasonable length of time. Private building sewerage systems must be discontinued and connections made to the public system when public sewers become available. All abandoned septic tanks and pits must have their contents removed and be filled with sand or gravel.

12-3 How is approval obtained for installing private domestic sewage treatment and disposal systems for industrial or public buildings and facilities?

Complete plans and specifications of the proposed facilities (such as trailer parks, industrial buildings, schools, theaters, hotels, apartment buildings, and similar constructions) must be submitted for

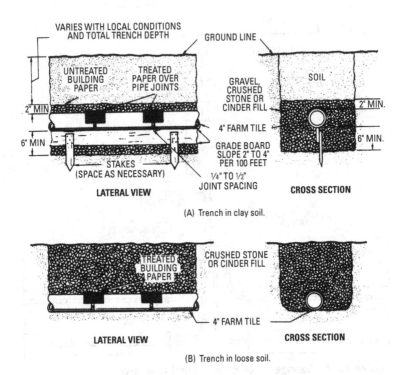

Figure 12-3 Trench layouts for various soil conditions.

approval to the local plumbing code administrative authority prior to advertising for bids and construction.

In addition to the plans and specifications, the following information must be provided:

- Plans of the septic tank and/or sewage disposal system
- Location of the water well (including distances to lakes or streams and to the property lot lines)
- Lot size including lot grade or slope
- Soil boring and soil percolation tests
- Proposed use and occupancy of the building or buildings

12-4 What supervision is required for septic tank installation soil percolation tests?

Soil boring and percolation tests shall be made under the direction and/or control of the approved plumbing code authority. Tests shall be conducted in a manner required by the code enforcement authority.

12-5 Do percolation test requirements and results vary?

Percolation test requirements and results do vary depending on locality, terrain, and soil conditions.

NOTE

Detailed soil maps are of value in determining estimated percolation rates and general soil characteristics. Soil maps are generally available from the area conservation office and/or area plumbing code administrative authority.

12-6 Are there specific septic tank installation location requirements?

Requirements vary depending on locality specifics and code requirements. An average or mean requirement for installation location would be as follows:

- Minimum distance from any building or building section shall be 5 feet.
- Soil absorption system shall be located not less than:
  - 25 feet from any building or dwelling
  - 55 feet from any reservoir, swimming pool, or well
  - 8 feet from any lot line
  - 30 feet from any water service
  - 55 feet from the high-water mark of any lake, stream, or waterway

12-7 Must septic tanks meet certain fabrication standards in their construction?

Generally septic tanks must meet the fabrication or construction standards of the local plumbing code administrative authority. The following requirements are generally included:

- All tanks shall be watertight and constitute an individual structure.
- Septic tanks shall be constructed of monolithic concrete, welded steel, or other approved materials.
- Cylinder tanks shall have an inside diameter of *not less* than 48 inches, and rectangular tanks shall have a minimum width of 36 inches with the longest dimensions parallel to flow direction.

- Precast concrete tanks shall have a minimum wall thickness of 2 inches, be marked to show liquid capacity, and have the registered trademark or name and address of the maker permanently inscribed on the tank near the outlet opening.
- Concrete tanks shall be designed to withstand usage pressures.
- Steel tanks shall be watertight, have structural stability, be made of new hot-rolled commercial steel, and be of a thickness as approved for the specific septic tank capacity size. Five-hundred to 1000-gallon steel tanks complete with their components must be constructed of a minimum steel thickness of 14-gage, and tanks of 2000 or more gallons in capacity of a minimum of 7-gage steel.

12-8 How is septic tank size determined?

Septic tank capacity is based on the number of persons using the building or on waste volume. Minimum liquid capacity of a septic tank is normally 500 gallons. Table 12-1 shows a typical sizing chart for one- and two-family residences.

Table 12-1 Septic Tank Capacity for One- and Two-Family Residences

Bedrooms	Typical Plumbing Fixtures	With Garbage Disposal Unit, Dishwasher, and Automatic Clothes Washer
2 or less	500 gallons	750 gallons
3	700 gallons	975 gallons
4	850 gallons	1200 gallons
5	950 gallons	1400 gallons
6	1150 gallons	1650 gallons

12-9 Is it permissible to install a septic tank within the interior foundation walls of a building?

Septic tanks are not permitted within the interior foundation walls of a building, nor shall a new building or an addition to an existing building be constructed within 5 feet of a septic tank.

12-10 Is a bedding base required when placing a septic tank?

All septic tank installations shall be provided with a minimum of (well-tamped) 3-inch thickness of gravel, sand, granite, or similar noncorrosive material. Materials shall not be above No. 14 standard size crushed material in size.

12-11 What types of backfill materials are recommended for septic tank installations?

Concrete tanks shall be backfilled with soil materials not exceeding 2 inches in size and well tamped. Backfill for steel tanks shall be gravel, sand, crushed limestone, or granite not exceeding No. 14 standard size material carefully tamped in place (size of materials should not exceed standard No. 14 stone and/or materials of such size that will pass a $1/2$-inch screen).

12-12 Are there recommended procedures for the disposal of effluent from septic tanks?

It is recommended that the effluent from septic tanks be disposed of by the soil absorption system, or in a manner approved by the local code administrative authority.

12-13 When should a private domestic sewage septic tank be cleaned?

Tanks should be cleaned when scum or sludge occupies one-third of the tank volume.

12-14 What procedures are used for septic tank sludge disposal?

Sludge from septic tanks is disposed of in accordance with the state and/or local code administrative authority (usually at a predetermined site), or by discharge into a public sewerage system. If these facilities are not available, the following disposal methods are generally recommended:

- By burial under 42 inches of earth at a minimum distance of 55 feet from a well or water source, and at least 550 feet from a place of habitation. It is further recommended that, in the burial location, there be 42 inches of soil between the buried sludge and the high ground water level.

- Some codes allow sludge to be spread on land not used for growing vegetables or for pasturing livestock, and at least 1200 feet from any stream, pond, flowage, lake, or place of habitation.

12-15 May sludge be disposed of in storm disposal runoff ditches?

Septic tank sludge disposal is not permitted in surface water disposal ditches and may not be discharged into any lake, stream, or flowage, or be buried within 70 feet of such locations.

12-16 What kind of trailer is considered to be of the residential type?

A residential trailer is a self-contained unit that is designed for the shelter of one or more people and comes complete with plumbing facilities and wheels (which allow the unit to be moved from one locality to another).

12-17 Which is considered the left side of the trailer?
The left side of the trailer is considered to be that section farthest from the curb when the unit is being towed.

12-18 Are there light and ventilation requirements for trailer toilet compartments?
General plumbing codes state that all trailer toilet compartments shall be provided with adequate light and ventilation facilities, particularly if permanently parked.

12-19 Is rodent- and vermin-proofing required for residential trailers?
All pipe and conduit openings through floors and walls shall be sealed with permanent materials attached so as to prevent entrance of vermin and rodents.

12-20 Does a requirement exist for proper slope and alignment of horizontal drainage piping and the proper securing of pipe and fixtures in residential trailers?
Horizontal drainage piping shall be installed in alignment at a uniform slope of not less than $1/8$ inch per foot. All piping shall be in good alignment, with provisions for contraction and expansion. In cold climates, fixtures shall be securely fastened and protected against freezing.

12-21 Is it permissible to install piping on the interior and/or exterior of trailers that would interfere with door or window operations?
Piping, fixtures, and equipment shall be installed so as not to interfere with normal operation of windows and doors.

12-22 May used materials be installed for residential trailer water supply systems?
Used materials are prohibited for water supply installation purposes in residential trailers.

12-23 What are the general recommendations for fixtures, fixture traps, and water closets to be installed in residential trailers?
Fixtures shall be of impervious materials of smooth finish installed in a manner to withstand road and travel shock and vibration. Fixture traps shall have not less than a 2-inch water seal located so as to prevent trap seal loss during normal use.
Water closets shall be of durable smooth finished materials installed in a manner so as to prevent spillage of trap seal contents

during travel, and also in a manner to prevent flushing of the water closet except when properly connected at a trailer camp or to an approved sewage disposal source.

When at trailer camps, connections between the trailer drainage system and the trailer park sewer system shall be made with a watertight, readily removable, local-code-approved flexible or semirigid connector.

Trailer water closets shall be provided with sufficient water supply to clean the interior of the unit when flushed, and also with a vacuum breaker so as to prevent possible potable water supply contamination. Vacuum breakers shall be of corrosion-resistant approved design and materials and shall be installed on the discharge side of the supply valve at a distance of not less than 6 inches above the fixture flood level.

12-24 Are there recommended locations for trailer drain outlets?

Residential trailer drain outlet termination pipe shall be visible from the road side of the trailer when in transit and be located at the rear of the wheel housing. Outlets shall be complete with permanently affixed tight closure caps or plugs. When in transit, trailer sewer and water terminals shall be capped.

12-25 What is the recommended trailer fixture branch drain slope in the vent stack area?

The branch drain in the stack area shall be sloped not more than $1/4$ inch per foot, and fixture branch connections at the stack shall be made with sanitary tees.

12-26 Are air gaps required for trailer plumbing fixtures?

An air gap is required for each fixture between the lowest opening from any water supply pipe or faucet and the flood-level rim of the fixture. For lavatories with effective openings not greater than $1/2$ inch in diameter, a minimum 1-inch air gap shall be provided. Air gaps shall be greater when effective opening sizes increase.

12-27 What is the minimum recommended size for trailer water service connection piping?

Minimum recommended size for domestic trailer water service connection pipe is $3/4$ inch. One-half-inch individual connection branches to each plumbing fixture are usually recommended.

12-28 Is there a minimum recommended distance between the water supply pipe and sewer connection pipe for trailers when at trailer parks?

Minimum recommended distances between trailer sewer and water pipe connections when permanently parked is 5 feet.

12-29 Some trailer parks provide hot-water facilities. If such hot-water piping is available, what is the minimum recommended water-heater storage capacity for residential type trailers?

Only approved automatic water heaters with minimum 5-gallon storage capacities are generally recommended for trailer use at trailer parks where hot-water piping is available for hookup.

12-30 Are safety devices required for trailer water heaters?

Water heaters for residential trailers shall be provided with a combination pressure and temperature relief valve approved by a nationally recognized testing association or laboratory. Relief valves shall be installed not more than 4 inches from the top of the water heater. Water heater tanks shall be subjected to not less than a 300-pound hydrostatic pressure test.

Tank outlets for draining shall be provided with shutoff valves and piped to permit draining to the outside of the trailer.

12-31 What is a sanitary sewer?

A *sanitary sewer* is a sewer that carries sewage but not storm water, surface water, or groundwater.

12-32 Describe a sewage seepage field.

A *seepage field* is an arrangement of perforated or open-joint piping buried underground that permits septic tank effluent to leach into the adjoining porous soil.

12-33 Describe a septic tank.

A *septic tank* is a watertight receptacle that receives raw sewage from a building or trailer sewer, allows for the digestion of organic matter, and permits liquid effluent to discharge into a holding pit or seepage field.

12-34 Describe a water well.

A *water well* is a driven, bored, or dug hole in the ground from which water is drawn or pumped.

12-35 Are there newer types of wastewater treatments for individual homes?

Newer technology provides for a cleaner fluid emission from the wastewater treatment unit. The septic tank has some limitations and some areas (especially heavily populated zones) no longer allow them to be installed.

One of the newer systems reduces normal household wastewater to clear odorless liquid in just 24 hours. It uses the same process as central treatment plants (see Figure 12-4).

The primary treatment compartment receives the household wastewater and holds it long enough to allow solid matter to settle

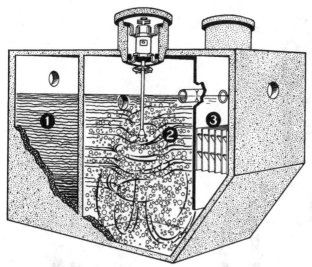

Figure 12-4 Home wastewater treatment plant. *(Courtesy Jet Inc.)*

to the sludge layer at the tank's bottom (see 1 in Figure 12-4). Organic solids are here broken down physically and biochemically by anaerobic bacteria (that is, those bacteria that live and work without oxygen). Grit and other untreatable materials are settled out and held back. The partially broken down, finely divided material that is passed on to the aeration compartment is much easier to treat than raw sewage.

In 2 of Figure 12-4, the aeration compartment takes the finely divided, pretreated material from the primary compartment and mixes it with activated sludge. It is then aerated. The aerator injects large quantities of fresh air into this compartment to mix the compartment's entire contents and to provide oxygen for the aerobic digester process.

The aerator (see Figure 12-5) is mounted in a concrete housing that rises to ground level to give it access to fresh outside air (see Figure 12-6). As air is injected into the liquid, the aerator breaks up this air into tiny bubbles so that more air comes in contact with the liquid, thus hastening the aerobic digestion process. Aerobic bacteria (which are bacteria that live and work in the presence of oxygen) then use the oxygen in the solution to break down the wastewater completely and convert it to odorless liquids and gases.

The final phase is shown in 3 of Figure 12-4. The settling/clarifying compartment has a tube settler that eliminates currents and encourages the settling of any remaining particulate material

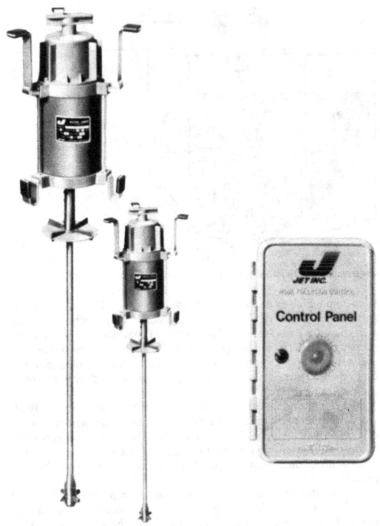

Figure 12-5 Electric motor-driven aerator and control panel.
(Courtesy Jet Inc.)

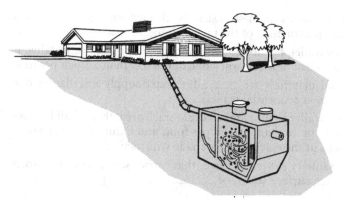

Figure 12-6 Treatment plant installed. *(Courtesy Jet Inc.)*

that is returned (by way of the tank's sloping end wall) to the aeration compartment for further treatment. A nonmechanical surface skimmer (operated by hydraulics) skims floating material from the surface of the settling compartment and returns it to the aeration compartment. The remaining odorless, clarified liquid flows into the final discharge line through the baffled outlet.

This type of home treatment system is often mandated by local health officials when there is a high water table or the soil has poor percolation.

Word Completion Exercises

Complete the following sentences by filling in the blanks. (See Appendix C for answers.)

1. All abandoned septic tanks and pits must have their contents removed and be filled with _____ or gravel.

2. Percolation test requirements and results do vary depending on locality, _____, and soil conditions.

3. Generally speaking, septic tanks must meet the fabrication or construction standards of the ____ plumbing code administrative authority.

4. Concrete tanks shall be backfilled with soil materials not exceeding 2 inches in size, and it shall be well ____.

5. A four-bedroom residence with a one-family occupancy will need a septic tank with a capacity of _____ gallons if there is no garbage disposal unit, dishwasher, and automatic clothes washer.

6. Septic tanks (domestic) shall be cleaned when scum or sludge occupies one-___ of the tank volume.

7. In a trailer, the horizontal drainage piping shall be installed in alignment at a uniform slope of not less than ___ inch per foot.

8. Used materials are _____ for water supply installation purposes in residential trailers.

9. In a trailer fixture branch drain stack area, there shall be slope at no more than ___ inch per foot, and fixture branch connections at the stack shall be made with sanitary tees.

10. A sanitary sewer is a sewer that carries sewage, but not storm water, surface water, or _____.

Chapter 13

Plastic Pipe and Fittings

Plastic pipe is used in many localities and communities for water supply, waste and vent piping, building drains and sewers, storm water drains, and other plumbing purposes.

The National Sanitation Foundation Testing Laboratory, Inc. (NSF), Ann Arbor, Michigan, has compiled and published a *Listing of Plastic Materials, Pipe, Fittings and Appurtenances for Potable Water and Waste Water*, and most plumbing codes require that the manufacturer, material, and trade name of approved plastic pipe and fittings be listed as specified in this publication. The use of plastic pipe for plumbing installations is subject to state and/or local unit plumbing code authority acceptance prior to installation.

The following questions and answers are typical of those asked in licensing examinations in reference to plastic plumbing pipe installations.

13-1 How shall joints in plastic piping be made?

Joints in plastic plumbing pipe shall be made only with approved fittings by either fusion-welded or solvent-welded connections, with metal clamps and screws of corrosion-resistant material and approved insert fittings, all in accordance with code approval authority.

13-2 What type of plastic pipe is recommended for aboveground interior building installation?

Plastic pipe and fittings for this purpose shall be chlorinated polyvinyl chloride (CPVC) with a minimum pressure of 200 psi at 73.5°F. Pipe and joints shall be installed in accordance with the manufacturer's recommendations and shall bear the NSF seal of approval.

Polyethylene (PE) plastic pipe shall be installed with insert and clamp-type fittings or thermal-welded fittings and joints. Polyvinyl chloride (PVC) pipe shall be installed with solvent-welded or flanged joints only. Polybutylene pipe (PB) shall be installed with insert-and-clamp, flange, or thermal-welded fittings and joints. All plastic insert fittings shall bear the NSF seal of approval.

13-3 What is the clearance requirement of plastic CPVC piping when placed adjacent to sources of excessive heat?

Plastic CPVC piping installed near, or parallel to, or crossing sources of excessive heat lines (such as heating devices or pipe, fume pipes, vent connectors, and similar piping) shall have not less than a minimum clearance of 6 inches. It is not recommended that plastic pipe be installed in any tunnel or boxed opening that contains uncirculated hot air, hot water, or steam.

13-4 What type of plastic pipe may be used for above- and below-ground purposes when used for building drains, waste, and vent piping?
Plastic pipe and fittings used for this purpose are as follows:

- Polyvinyl chloride (PVC) meeting standards CS-207-60 or CS-272-65 or ASTM D-2665-68 or ASTM D-1785-67a
- Acrylonitrile-butadiene-styrene (ABS) (not generally approved for acid wastes) meeting standards ASTM D-2661-67 or CS-270-65 or of the latest approved edition

13-5 What types of supports are required for plastic pipe installations?
Horizontal plastic pipe supports should not exceed 3-foot spacing. Vertical pipe supports should not exceed 4-foot spacing with supports also at the base.

13-6 Are expansion joints and fittings recommended when plastic piping is used for plumbing purposes?
PVC plastic pipe has a considerably greater linear coefficient of thermal expansion than conventional metallic pipeline materials. Provided that this fact is recognized and allowances are made for expansion, the expansion factor does not prove to be troublesome. Expansion provision is recommended for each 35 feet of installed straight horizontal and vertical pipe runs.

Table 13-1 shows the change in length of PVC pipelines in inches for various temperatures. From these figures, calculations can be

Table 13-1   Change in Length (Inches) of Vinyl DWV for Changes in Operating Temperature

Pipe Length	Temperature Range			
	10°F	30°F	50°F	100°F
1 inch	0.0036 inch	0.0108 inch	0.018 inch	0.036 inch
2 inches	0.0072 inch	0.0216 inch	0.036 inch	0.072 inch
5 inches	0.0180 inch	0.0540 inch	0.090 inch	0.180 inch
10 inches	0.0360 inch	0.1080 inch	0.180 inch	0.360 inch
20 inches	0.0720 inch	0.2160 inch	0.360 inch	0.720 inch

made of the expansion likely to be encountered over the operating range of PVC piping. (Dimensions given in Table 13-1 are only the theoretical expansion criteria for PVC piping.)

13-7 What are the underground installation procedures for plastic piping?

To ensure good underground installation permanence for plastic pipe, the following procedures are recommended:

1. *Preparation of trench*
   a. The bottom of the trench should be sloped in drainage applications to ensure adequate pipe pitch (minimum of $\frac{1}{8}$ inch per foot).
   b. A bed of sand or pea rock should be provided in the trench bottom for the installed pipe to rest on a 3-inch bed for up to 4-inch diameter pipe, or a 4-inch bed for up to 8-inch diameter pipe.

2. *Initial backfill*
   a. The initial backfill must be free of large rocks and other foreign materials likely to damage piping.
   b. The initial backfill must be carefully tamped around pipe.

3. *Final backfill*
   a. The final backfill should be free of large rocks and foreign materials.
   b. The fill (dirt) should be heaped over excavation to allow for settling.

4. *Under-concrete applications*
   a. Bury pipe in sand fill under concrete.
   b. Insulate all risers through concrete slab to allow for expansion and/or contraction.

NOTE

Always water-test or air-test piping for leaks before pouring the concrete slab.

13-8 How is plastic pipe cut and assembled?

Typical cutting and assembling procedures for plumbing plastic pipe installations are shown in Figure 13-1 (pipe cutting), Figure 13-2 (cleaning out pipe), Figure 13-3 (applying adhesive), and Figure 13-4 (assembling prepared pipe and fitting).

Figure 13-1 Method of cutting plastic pipe. Plastic pipe can be cut with a saw or tubing cutter.

13-9 Describe the plastic pipe terms CPVC and PB.

CPVC plastic pipe is chlorinated polyvinyl chloride plastic piping. PB is polybutylene plastic piping (that is, thermoplastic flexible piping). Both CPVC and PB piping are ordinarily acceptable in all localities. Check your local code requirements. CPVC conforms to the American Society for Testing and Material (ASTM) D-2846. PB conforms to ASTM D-3309. Both are generally accepted by the Federal Housing Administration. Figure 13-5 shows typical plastic water supply fittings. Figure 13-6 shows typical plastic drain-waste and vent fittings. Figure 13-7 illustrates a typical CPVC pipe water heater/bathtub hookup.

13-10 Describe a building septic system.

A building septic system includes a sewer line leading from the building, a septic tank, and a seepage field. The septic tank is the location where solid and liquid building wastes decompose by bacterial action. Solids settle while the effluent flows into the

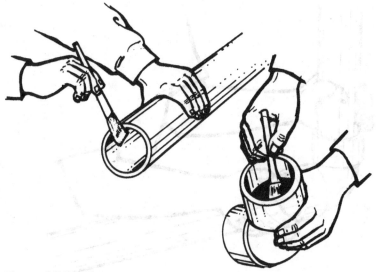

Figure 13-2 Cleaning both pipe and connectors.

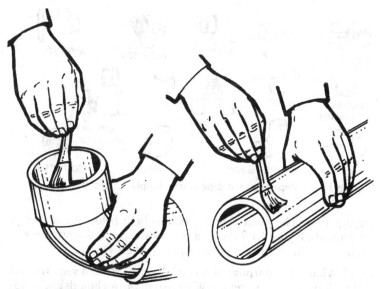

Figure 13-3 Applying welding adhesive.

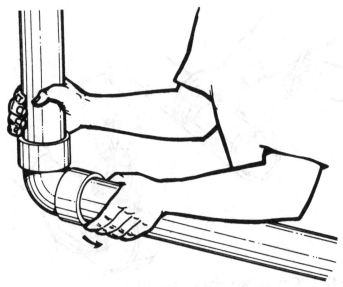

Figure 13-4 Assembling pipe and connectors.

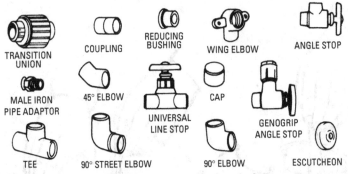

Figure 13-5 Typical plastic pipe water supply fittings. *(Courtesy Genova, Inc.)*

seepage field. The entire septic system is built below grade (below the ground). Figure 13-8 shows a typical seepage bed septic system. Figure 13-9 illustrates a typical seepage bed cross section.

13-11 What is the purpose of a vinyl drain-waste and vent system?

The drain-waste and vent (DWV) system carries liquids and solids out of the building. Figure 13-10 shows a typical vinyl drain-waste and vent system for a bathroom and basement sump pump.

Figure 13-6 Typical plastic pipe drain-waste and vent fittings.
(Courtesy Genova, Inc.)

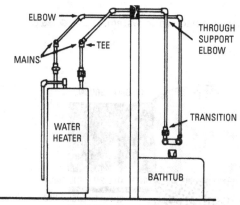

Figure 13-7 A typical CPVC pipe water heater/bathtub hookup.
(Courtesy Genova, Inc.)

13-12 What equipment is generally required for rigid vinyl installation?

The following listed equipment is the minimum required for DWV installation procedures:

- Welding adhesive and cleaning compound.
- Clean natural bristle brush (two required). Width should be half the diameter of pipe being installed.

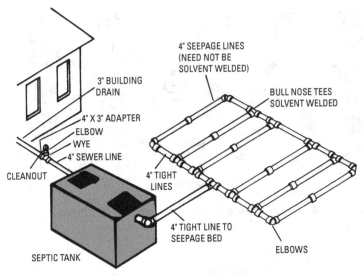

Figure 13-8 A typical seepage bed septic system. *(Courtesy Genova, Inc.)*

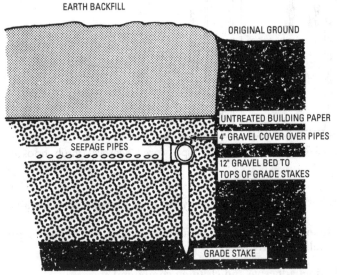

Figure 13-9 A typical seepage bed cross section. *(Courtesy Genova, Inc.)*

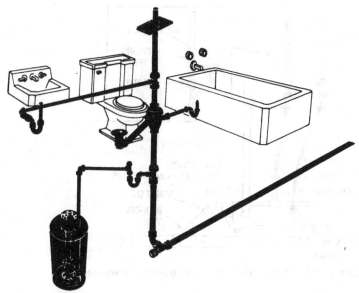

Figure 13-10 A typical vinyl pipe drain-waste and vent system.
(Courtesy Genova, Inc.)

- Miter box and handsaw.
- File and/or sandpaper.
- Clean cloth.
- Wood support (2 inch × 4 inch) to hold pipe end off ground during adhesive welding.
- Strap wrench.

NOTE

Extreme care should be used to prevent solvent welding compounds from spilling and damaging finished floors.

13-13 Is plastic piping used for kitchen and washer installations?

Plastic piping is used for washer and kitchen plumbing piping installations as approved by administrative code authority. Figure 13-11 illustrates a typical plastic drainage pipe layout for a washer. Figure 13-12 shows a typical plastic pipe drainage system for a kitchen layout.

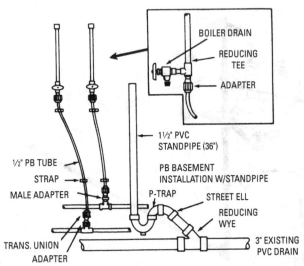

Figure 13-11 A typical plastic pipe layout for a washer.
(Courtesy Genova, Inc.)

13-14 Are there any special requirements for unloading and handling plastic pipe?

A level, flat, clean site free of nails, rocks, and other debris that might damage the pipe is preferable. Wooden supports should be provided at not more than 3-foot intervals in locations where debris or other items exist (see Figure 13-13).

In locations where mechanical handling is used, metal slings and hooks should be avoided. Standard pipe rope slings as illustrated in Figure 13-14 are generally recommended.

In storage racks, pipe shall have at least six evenly distributed supports for each length of pipe (see Figure 13-15).

13-15 Are there special workmanship requirements for plastic pipe installations?

Good workmanship is necessary for plastic pipe installation, as it is for the installation of other types of plumbing materials including copper and cast-iron piping. A job can be successfully completed only by following instructions and taking pride in the work done.

Multiple-Choice Exercises

Select the right answer and blacken *a*, *b*, *c*, or *d* on your practice answer sheet. (See Appendix C for answers.)

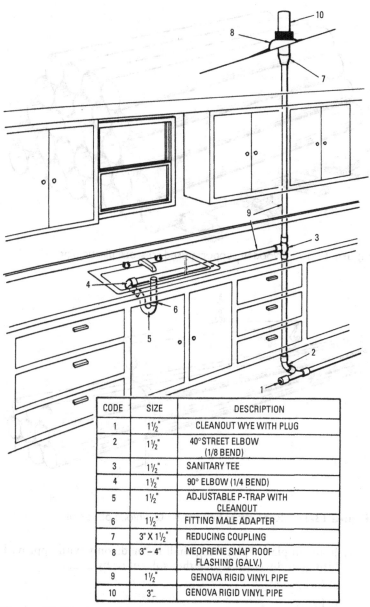

CODE	SIZE	DESCRIPTION
1	1½"	CLEANOUT WYE WITH PLUG
2	1½"	40°STREET ELBOW (1/8 BEND)
3	1½"	SANITARY TEE
4	1½"	90° ELBOW (1/4 BEND)
5	1½"	ADJUSTABLE P-TRAP WITH CLEANOUT
6	1½"	FITTING MALE ADAPTER
7	3" X 1½"	REDUCING COUPLING
8	3" – 4"	NEOPRENE SNAP ROOF FLASHING (GALV.)
9	1½"	GENOVA RIGID VINYL PIPE
10	3"	GENOVA RIGID VINYL PIPE

Figure 13-12 A typical plastic pipe drainage system in a kitchen.

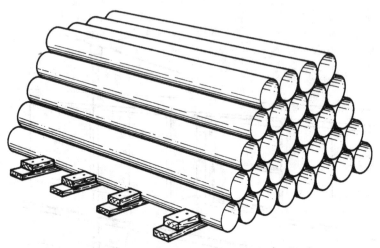

Figure 13-13 Wooden supports used to store plastic pipe.

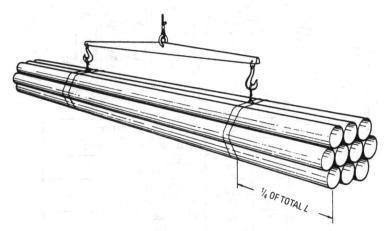

¼ OF TOTAL *l*

Figure 13-14 Rope slings being used to unload plastic pipe.

1. Joints in plastic plumbing shall be made only with approved fittings and welded with either solvent or by ＿＿＿.

   a. fusion
   b. brass
   c. copper
   d. solder

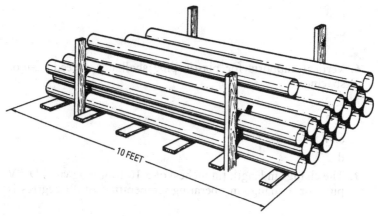

Figure 13-15 Typical plastic pipe storage racks.

2. Plastic pipe for aboveground interior building installations shall *not* be _____.

   a. CPVC

   b. PVC

   c. PB

   d. Lead

3. It is not recommended that plastic pipe be installed in any _____ or boxed opening that contains uncirculated hot air, hot water, or steam.

   a. tunnel

   b. area

   c. location

   d. place

4. The type of plastic pipe that may be used for above- and below-ground purposes for drains, waste, and vent piping is called _____.

   a. CPVC

   b. DPVC

   c. PVC

   d. ABS

5. Horizontal supports for plastic pipe shall not exceed ___-foot spacing.

   a. 2

b. 3

c. 4

d. 5

6. Vertical supports for plastic pipe shall not exceed __-foot spacing.

   a. 2

   b. 3

   c. 4

   d. 5

7. The change in length (in inches) of a 10-foot long vinyl DWV pipe for a change in operating temperature of 30 degrees is _____.

   a. 0.036

   b. 0.0036

   c. 0.108

   d. 0.360

8. The description for CPVC pipe is _____.

   a. chlorinated polyvinyl chloride plastic

   b. polybutylene plastic piping

   c. chlorinated plastic container

   d. chlorinated polyvinyl chloride

9. The drain-waste and vent system carries liquids and solids out of the building. A typical vinyl drain-waste and vent system is referred to as a _____.

   a. DWV system

   b. drain-way vent system

   c. downward vent system

   d. dominant drain vertical system

10. Plastic should be unloaded and handled with care. It can be damaged by rocks and debris on a building site. When in storage racks, the pipe shall have at least __ evenly distributed supports for each length of pipe.

    a. 5

    b. 6

    c. 4

    d. 3

Chapter 14

Plumbing Installation Inspection Tests

Piping for plumbing installation must be watertight and airtight. Following are three kinds of tests for determining tightness:

- Water (hydrostatic)
- Air
- Smoke

On most job sites (particularly on large installations), water tests are applied progressively as the work proceeds to avoid the increased expense in case it is necessary to remove defective parts. Earthen drains should be carefully tested for leakage before the trenches are filled.

On water tests, the low end of the line is plugged and filled with water. A pressure of at least 1 pound per square inch should be applied. A leak, if not visible, will be indicated by the water settling in the pipe and should be found and made tight. The water test should be applied to all soil, waste, and vent pipe in the building. In making this test, plug the building drain where it passes out of the building, plug the fresh-air inlet, and plug all other outlets up to the highest opening.

The plugs used to close the openings are known as *proving* or *testing plugs* and are shown in Figure 14-1. If the branch pipes

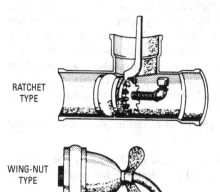

Figure 14-1 Typical proving or testing plugs.

RATCHET TYPE

WING-NUT TYPE

are lead, they can be closed by soldering caps over the ends (that is, assuming the outlets have been tightly closed by the plugs). If the water level falls after the pipes are filled with water, a leak is indicated. If the water remains at the same level, the system is tight. In looking for leaks, first inspect all the plugged openings to see if they are tight. If the weather is too cold to apply the water test, the air test may be substituted.

In making an air test, air is pumped into the system (after closing all of the openings with testing plugs) to a pressure of 10 pounds per square inch as indicated on a mercury gage (rather than a spring gage). A leak is indicated by a fall in the level of the mercury column. To detect a leak, apply soapy water with a brush; if there is a leak, bubbles will form.

Generally the air test gives a practically uniform pressure over the entire system, whereas with the water test, the pressure is greatest at the lowest level and least at the highest level. Caution should be used with the water test. Do *not* apply it where the height of the vertical pipes is so great as to produce pressure too great for the pipes.

In making the smoke test, the traps are properly filled with water and all fixtures set. The opening at the fresh-air inlet should be closed. The smoke machine or apparatus is generally connected to the fresh-air inlet or to one of the extensions above the roof. Smoke is placed in the machine and forced into the system until the pressure equals 1 or 1½ inches of water. If the machine is applied to the extensions above the roof, the smoke should be pumped in until it comes from the fresh-air inlet, after which the inlet should be closed. Leaks are detected by smell or, if the smoke becomes dense enough, smoke can be seen escaping. When the smoke or air test is made, the applied pressure should be about 1½ inches of water after the fixtures are installed. If any of the traps blow through at a lower pressure than 1½ inches, they should be readjusted until they will hold that pressure. If that cannot be accomplished, a better trap should be installed. After the pressure is put on the system, closely watch the water gage for several hours. If the water column falls, it indicates a leak in the system.

The following are typical questions asked in plumbing licensing examinations.

14-1 Are plumbing code inspections conducted to verify good workmanship and satisfactory material installations for newly installed plumbing systems?

General plumbing code requirements state that all piping of a drainage or plumbing system in cities and villages that have local plumbing codes must be tested by the licensed plumber in charge of the work. A certificate in writing (showing the test results as to water- and airtightness of the system installed) shall be presented to the owner, or as required by the local plumbing administrative authority.

14-2 How are tests of drainage and vent systems accomplished?

Three types of tests are generally used. These include the water test, air test, and smoke test.

If a water test is made, the test is applied to the sanitary plumbing system inside the building by sections or in its entirety. The water test is conducted by closing all openings in the pipe with approved testing plugs to the highest opening above the roof and then filling the system completely with water and noting if leaks exist.

An air test is made by attaching a test apparatus or air compressor to a suitable opening after other openings have been closed off and then forcing air into the sanitary system until there is a uniform pressure in the system to balance a mercury column 10 inches high or a pressure of 5 pounds per square inch on the entire sanitary system for at least a 12-minute period of time. Pipes must be found free from defects and leaking joints.

Normally, buildings four or more stories high are tested in sections. Smoke tests are usually used in sanitary pipe systems if there is a possibility that the system has become defective. After all system openings have been properly sealed, a dense smoke is forced under pressure into the system equivalent to a 1-inch water column for at least a 12-minute period with visual tests for smoke leaks during this period.

14-3 Is it necessary to air-test or water-test rain conductor pipe installed on the interior of the building?

Water or air tests shall be applied to all rain conductor piping, including their roof connections *if they are installed within the walls of the building.*

14-4 If tests or inspections indicate defective materials, must these materials be replaced?

If tests reveal poor workmanship or defective materials, the work and defective materials shall be removed and/or replaced, and work shall be retested so as to conform to the area plumbing code requirements.

It is the duty of the plumber holding the permit to make sure that the installation is in accordance with locality plumbing code requirements.

14-5 Is it necessary to test the water supply system?

Water supply systems shall be tested for tightness under a water pressure of not less than 50 percent greater than the proposed-use water pressure. Only approved potable water shall be used for this test.

14-6 Are there plumbing system maintenance requirements?

Building plumbing and drainage systems shall be maintained in a safe, sanitary operating condition at all times.

14-7 How are plumbing permits obtained?

Applications for plumbing permits are made to the area code administrative authorities in accordance with their requirements.

14-8 Are permits for plumbing repairs required?

Repairs that include only removal of stoppages, repair of leaks, or replacements of defective parts or fixtures do not normally require a plumbing permit, provided no extensive piping changes are made. In all instances, the potable water supply system shall be protected at all times against contamination.

14-9 Do authorized plumbing inspectors have a right to enter private buildings for purposes of plumbing inspections?

Under normal circumstances and at reasonable times, approved plumbing inspectors do have the authority to enter public and private buildings being remodeled or repaired, or those under construction where plumbing systems are involved. Inspection procedures and authority in accordance with local administrative plumbing code requirements are required.

14-10 Is it permissible for private citizens to install plumbing in their own residences?

Normally, bona fide owners can personally install plumbing in their own residences, provided they apply for and secure a permit, perform their own work in accordance with the local plumbing code, and receive a work inspection that is approved (on completion) by the plumbing code inspector.

Multiple-Choice Exercises

Select the right answer and blacken a, b, c, or d on your practice answer sheet. (See Appendix C for answers.)

1. Piping for plumbing installation must be watertight and airtight. The three kinds of tests for determining tightness are water, air, and _____.

a. smoke

b. electronic

c. mechanical

d. hydro

2. The water test should be applied to all soil, _____, and vent pipe in the building.

a. waste

b. drain

c. water

d. hot water

3. The person who conducts the tests on the plumbing system in a new installation is _____.

a. the apprentice

b. the licensed plumber on the job

c. the city inspector

d. the building contractor

4. The water test is conducted by closing all openings in the pipe with approved testing _____ to the highest opening above the roof and then filling the system completely with water and noting if leaks exist.

a. masking tape

b. rubber caps

c. plugs

d. plastic tape

5. If installed within the walls of the building, it is necessary to air- or _____-test rain conductor pipe.

a. smoke

b. electronically

c. mechanically

d. water

6. It is the _____ of the plumber holding the permit to make sure that the installation is in accordance with locality plumbing code requirements.

a. responsibility

b. duty

    c. job

    d. right

7. Water supply systems shall be tested for tightness under water pressure of not less than 50 percent greater than the proposed-use water pressure. Only approved _____ water shall be used for this test.

    a. bottled

    b. lake

    c. potable

    d. river

8. Plumbing permits are usually obtained by making _____ to the area code administrative authorities.

    a. an appeal

    b. an application

    c. a trip

    d. a call

9. Permits are not required to repair leaks, remove stoppages, or replace defective parts or fixtures. In all instances, the ___ water supply system shall be protected from contamination.

    a. fresh

    b. test

    c. potable

    d. drainage

10. Normally, bona fide _____ can personally install plumbing in their own residences, provided they apply for and secure a permit, perform their own work in accordance with the local plumbing code, and receive a work inspection that is approved (on completion) by the plumbing code inspector.

    a. renters

    b. lessees

    c. owners

    d. banks

Chapter 15

Steam and Hot-Water Heating

Design and installation questions regarding steam and hot-water heating are not always asked in state plumbing licensing examinations. However, at times, reference is made to steam and hot-water heating piping systems. Therefore, this chapter on steam and hot-water heating will briefly discuss both types of heating systems.

Steam Heating

Steam is an effective heating medium and is adaptable to almost any type of building. One of the simplest steam-heating systems is the one-pipe gravity type shown in Figure 15-1. Installations of

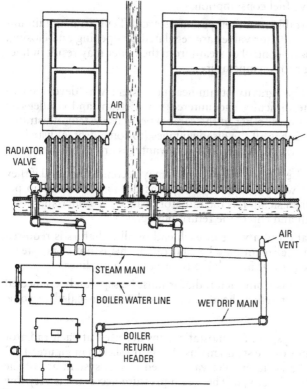

Figure 15-1 A one-pipe gravity steam-heating system.

(Courtesy Dunham-Bush, Inc.)

this type are generally limited to moderate-sized buildings where the radiators can be positioned at least 24 inches above the water level in the boiler. This type of installation is simple to operate and the initial cost is low.

Some inherent disadvantages of the one-pipe gravity steam-heating systems are as follows:

- Large piping and radiator valves are required to allow the condensate to return against the resistance offered by the steam flow.
- Steam and condensate flow in opposite directions with a possibility of water hammer.
- Air valves are required. Failure of these valves to always open to allow the escape of air may result in slow heat buildup and excessive fuel consumption.
- Comfortable room temperatures are difficult to maintain unless the radiator valves are regulated by opening and closing. Automatic control of steam from the boiler may result in fluctuating room temperatures.

The *two-pipe* gravity steam-heating system was developed to overcome the difficulty encountered when steam and condensate flow against one another in the same pipe. This system, illustrated in Figure 15-2, has all the disadvantages (to some degree) listed for the one-pipe system. Additional disadvantages are as follows:

- The midsections of the radiators may become air-bound if they are not water-sealed. This happens during the warm-up period when steam fills the radiator nearest the boiler and flows through and enters the return piping.
- Installation of a valve at each end of all radiators is required so that the steam may be shut off. If this is not done, steam may be present in both the return and supply lines.
- The returns from each radiator must be separately connected into a wet return header, or must be water-sealed in some manner.

In a *vapor system*, thermostatic traps are used at each radiator and at the ends of the steam mains. These are shown in Figures 15-3 and 15-4. The radiator inlet valves used in this system are of the graduated or orifice type. The steam pressure necessary is very low (often less than one pound). This type of installation can be used

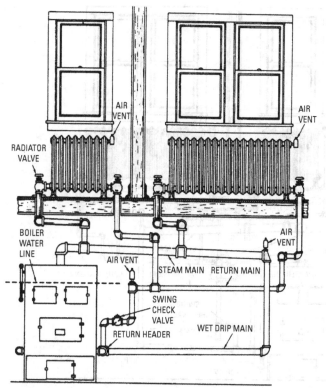

Figure 15-2 A typical two-pipe gravity steam-heating system.
(Courtesy Dunham-Bush, Inc.)

in buildings where 24 inches or more can be provided between the boiler water level and the end of the return line.

The following advantages should be considered:

- Air cannot enter the system as it is closed. Thus, a moderate vacuum is created by the condensing steam producing steam at lower temperatures.
- An even, quiet circulation of steam is provided with no air-binding or noisy water hammer.
- Room temperatures can be closely and automatically regulated by thermostatic controls.
- Radiator air valves are not required.

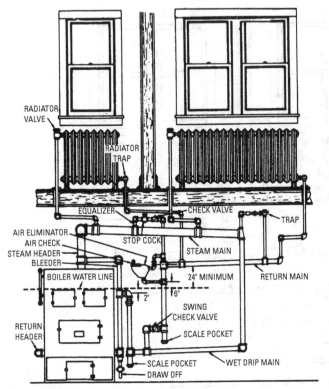

Figure 15-3 A vapor steam-heating system. *(Courtesy Dunham-Bush, Inc.)*

Disadvantages of a vapor steam-heating system include the following:

- Comparatively large pipe sizes are necessary.
- Only low steam pressure is possible.
- The condensate must return to the boiler by gravity.

The condensate must return to the boiler by gravity in this type of heating system and may back up in the vertical return pipe when there is excess steam pressure in the boiler. As a result of this, an air eliminator must be installed well above the level of the water in the boiler, yet low enough for it to close before the return water is of sufficiently high level to enter the return main. Close control of the boiler pressure is required in this type of system.

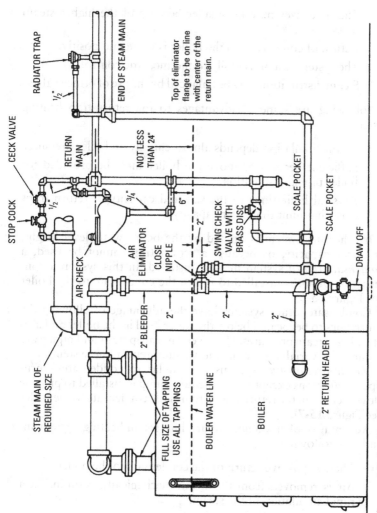

Figure 15-4 Pipe Installation of a vapor system. *(Courtesy Dunham-Bush, Inc.)*

Labels in figure:
- RADIATOR TRAP
- ½"
- END OF STEAM MAIN
- Top of eliminator flange to be on line with center of the return main.
- CECK VALVE
- STOP COCK
- RETURN MAIN
- ½"
- NOT LESS THAN 24"
- ¾"
- 6"
- AIR CHECK
- AIR ELIMINATOR
- CLOSE NIPPLE
- 2"
- SWING CHECK VALVE WITH BRASS DISC
- SCALE POCKET
- SCALE POCKET
- DRAW OFF
- STEAM MAIN OF REQUIRED SIZE
- 2" BLEEDER
- 2"
- 2"
- 2"
- 2" RETURN HEADER
- FULL SIZE OF TAPPING USE ALL TAPPINGS
- BOILER WATER LINE
- BOILER

A *return-trap* heating system closely resembles the vapor system except that the return trap provides a positive return of condensate to the boiler (see Figure 15-5). This type can be used in all but the largest buildings if the equivalent direct radiation (EDR) capacity is not greater than the return-trap capacity.

Advantages of a return-trap steam-heating system include the following:

- The pipe sizes may be smaller because of the higher steam pressures.
- Return of condensate to the boiler is rapid and positive.
- The system responds easily to thermostatic control.
- Steam distribution may be balanced by the use of orifice valves.

Following are some disadvantages to the return-trap heating system:

- Steam circulation depends almost entirely on boiler pressure.
- Sufficient headroom above the boiler must be provided for boiler piping installation.
- Physical limitations to the size and capacity of return traps exist that limit the boiler capacity.

If a heating system is limited by the boiler water-level height, by boiler capacity, or because a return trap cannot be used, a *condensate-return* system may be installed. In this system a condensation pump is installed to return the condensate to the boiler (see Figure 15-6).

Condensate-return systems have the advantage of allowing the return lines to be located below the water level in the boiler and also allow high steam pressures. However, large drip traps and piping are required for variable-vacuum and vacuum-return-line systems.

Vacuum-return-line systems are similar to condensate-return-type installations except that a vacuum pump is installed to provide a low vacuum in the return line to return the condensate to the boiler (see Figure 15-7).

Advantages of a vacuum-return-line steam-heating system include the following:

- There is positive return of the condensate to the boiler.
- Air is removed from the steam mechanically, resulting in a rapid circulation of the steam.

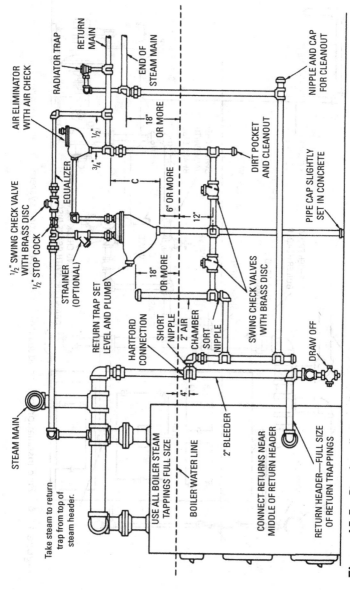

Figure 15-5 Pipe Installation of a return-trap steam-heating system. *(Courtesy Dunham-Bush, Inc.)*

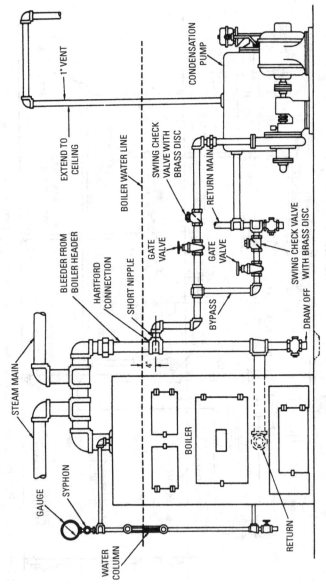

Figure 15-6 Pipe installation of a condensate-return–type heating system. *(Courtesy Dunham-Bush, Inc.)*

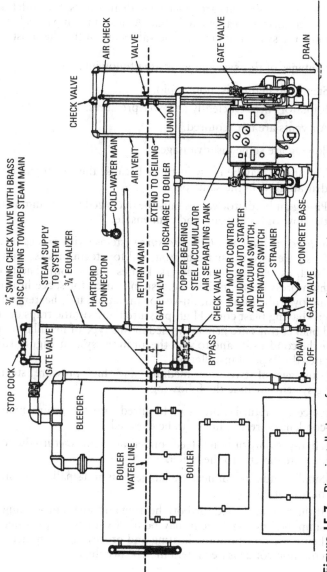

STOP COCK

¾" SWING CHECK VALVE WITH BRASS
DISC OPENING TOWARD STEAM MAIN

STEAM SUPPLY
TO SYSTEM

¾" EQUALIZER

CHECK VALVE

AIR CHECK

VALVE

GATE VALVE

COLD-WATER MAIN

AIR VENT

UNION

HARTFORD
CONNECTION

GATE VALVE

RETURN MAIN

EXTEND TO CEILING

DISCHARGE TO BOILER

COPPER BEARING
STEEL ACCUMULATOR
AIR SEPARATING TANK

CHECK VALVE

PUMP MOTOR CONTROL
INCLUDING AUTO STARTER
AND VACUUM SWITCH,
ALTERNATOR SWITCH

STRAINER

CONCRETE BASE

GATE VALVE

DRAIN

BLEEDER

GATE VALVE

BYPASS

DRAW
OFF

BOILER
WATER LINE

BOILER

Figure 15-7 Pipe installation of a vacuum-return-line steam-heating system. *(Courtesy Dunham-Bush, Inc.)*

261

- Smaller pipe sizes may be used because of the greater pressure differential between the supply and return lines.

The *vacuum-air-line* is a variation of the one-pipe steam-heating system. Radiator air vents are replaced with air valves, the outlets of which are connected to a return air line. A vacuum pump is included to exhaust the system of air. Air line valves are of the thermostatic type.

Advantages of this system include the following:

- Steam circulation is more rapid.
- Radiators heat efficiently at lower pressures.
- Air vents are not required.

Following are some disadvantages:

- The steam may be noisy at times because the steam and condensate flow in the same pipe.
- Piping and radiator valves must be oversized to accommodate the flow of both steam and condensate.

Two-pipe medium and *high-pressure* systems (with ranges of 25 to 125 psi) are used for space heating and for steam-process equipment (such as water heaters, dryers, kettles, and so on). The two-pipe, high-pressure steam-heating system is used for space heating and employs tube radiation, fin-vector radiation, unit heaters, and fan units having blast coils. High-pressure thermostatic traps or inverted bucket traps are generally used in this type of installation to handle the condensate and to vent the air in the system. Figure 15-8 illustrates this type of installation.

Advantages of two-pipe medium and high-pressure steam-heating systems include the following:

- Since the return water can be lifted into the return mains, condensate-return lines can be elevated.
- High-pressure or boiler-feed condensate pumps return the condensate directly to the boiler.
- Smaller radiators or heat-exchange units and smaller pipe sizes may be used.
- If the system is used for both heating and steam processing, a simplified piping system may be used with the supply mains being used for both the heating and processing equipment. A common condensate system may be installed.

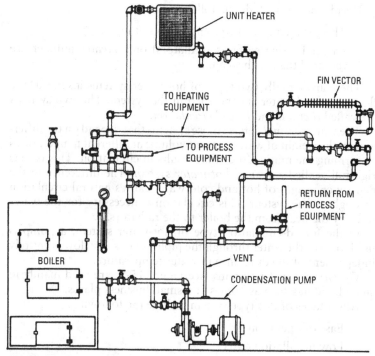

Figure 15-8 A typical two-pipe, high-pressure steam-heating system.
(Courtesy Dunham-Bush, Inc.)

Hot-Water Heating

Hot-water heating systems transmit only sensible heat to radiators, as distinguished from steam systems that heat principally by the latent heat of evaporation. The result is that the temperature of the radiators of a steam system is relatively high compared to those of a hot-water system. In a hot-water system, latent heat is not given off to a great degree, so more heating surface is required.

Advantages of hot water heating include the following:

- Temperatures may widely vary, so that it is more flexible than low-pressure (above atmospheric) steam systems.
- The radiators will remain warm a considerable time after the heat-generating fire has gone out. Thus, the system is a reservoir for storing heat.

Disadvantages include the following:

- There is a danger of freezing when not in use.
- Larger heating surfaces (radiators) or a greater number are required than with steam.

There are actually two types of hot-water systems insofar as the flow of heated water is concerned in the system. The two systems are called *thermal* and *forced circulation*.

The word *thermal* defines those systems that depend on the difference in the weight of water per unit volume at different temperatures as forming the motive force that results in circulation. This type is rightfully called a *gravity hot-water system*. The difference in the density or weight of hot and cold water causes natural circulation throughout the system. This circulation is necessary for the water to carry the heat from the boiler to the radiators.

In the forced-circulation type of hot-water system, a pump is used to force the water through the piping. Thus, the flow is entirely independent of the difference in water temperature.

Gravity hot-water systems (see Figure 15-9) are used mainly in small buildings such as homes and small business places.

Advantages of this type system include the following:

- Ease of operation
- Low installation costs
- Low maintenance costs

Disadvantages include the following:

- Possible water damage in case of leaks.
- Rapid temperature changes result in a slow response from the system.
- Properly balancing the flow of water to radiators is sometimes difficult.
- Nonattendance when the heat-generating unit fails may result in a freeze-up.
- Flow depends on gravity and, as a result, larger pipe sizes are required for good operation.

Forced hot-water heating systems require a pump that forces the water through the piping system. Limitations of flow (dependent on water-temperature differences), do not exist in this type of system. This kind of system may be of either the one- or two-pipe variety. In two-pipe systems, either direct or reversed returns and either upfeed

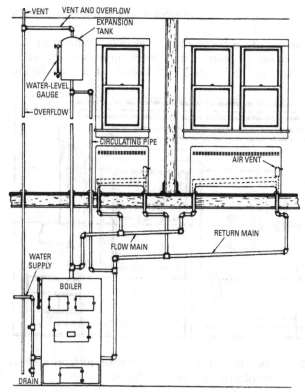

Figure 15-9 A typical gravity-pipe hot-water heating system.
(Courtesy Dunham-Bush, Inc.)

or downfeed mains may be used. The path of the water from the boiler into and through the radiator and back again to the boiler is almost the same length for each of the radiators in the system. It is common to use one-pipe forced-circulation systems for small- and medium-sized buildings when hot water is used as the heating medium.

Advantages of forced circulation include the following:

• There is rapid response to temperature changes.

• Smaller pipe sizes may be installed.

• Room temperatures can be automatically controlled if either the burner or the flow of water is thermostatically controlled.

• There is less danger of water freeze and damage.

A major disadvantage is the requirement for venting at all high points.

Radiant-panel heating is the method of heating a room by raising the temperature of one or more of its interior surfaces (floor, walls, or ceiling), as illustrated in Figure 15-10, instead of heating the air.

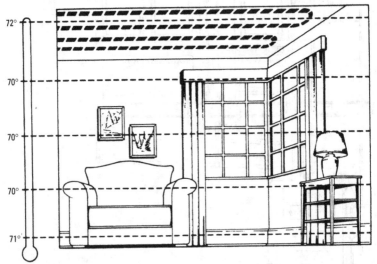

Figure 15-10 Radiant heating warms the interior surfaces such as ceiling, walls, and floor of a room instead of the air.
(Courtesy Chase-Brass and Copper Co.)

One of the most common methods of achieving radiant heating is by the installation of specially constructed pipe coils or lengths of tubing in the floor, walls, or ceiling. These coils generally consist of small-bore wrought-iron, steel, brass, or copper pipe, usually with an inside diameter of $3/8$ to 1 inch. Every consideration should be given to complete building insulation when radiant-panel heating is used.

Air venting is necessary in the proper control of any panel hot-water heating system. A collection of air in either the circuit pipe or pipe coils results in a shortage of heat. Figure 15-11 illustrates an installation arrangement that will permit uniform venting. Because of the continuous slope to the coil connections, it may be sufficient to install automatic vents at the top of the return riser only, omitting such vents on the supply riser.

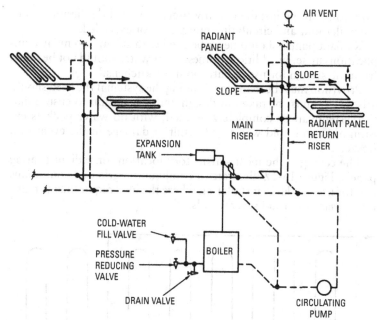

Figure 15-11 Diagram of a radiant-panel heating system showing the draining and venting. *(Courtesy Chase-Brass and Copper Co.)*

The following are some advantages of radiant-panel heating:

- Radiant-panel heating eliminates radiators and grills, thus providing more floor space and resulting in better furniture arrangements and wall decorations.
- There is less streaking of walls and ceiling because of lower velocities of air currents.
- It provides warm floors in homes that do not have basements.
- It simplifies interior architectural and engineering building designs.
- A well-designed and -installed radiant-panel heating system provides low operating and maintenance costs.

Hot-water radiant panels can be installed in nearly any type of building, with or without a basement or excavated section. Conventional hot water boilers are used. Units of this type are available in compact types that fit into small spaces and are fired by gas or oil.

Additions can be installed at any time, provided the limitations of the boiler unit and circulating pump are not exceeded.

Radiant panels should never be used with steam; too many complications arise. In addition, domestic hot water should not be taken from the system for use in bathrooms or kitchens.

Provision must be made for draining the system if the need should arise. Care must be taken in design and installation to ensure that the system can be completely drained, with no water pockets existing that will hold water and result in damage in the event of a freeze.

The ceiling is the most satisfactory location for radiant heating panels. Figure 15-12 illustrates a combination grid and continuous coil for use in a large installation where the heat requirements make it necessary to install several coils.

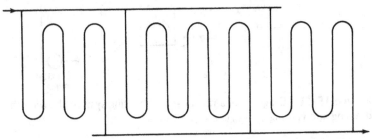

Figure 15-12 A combination grid and continuous coil radiant heating panel as used in a large installation. (Courtesy Chase-Brass and Copper Co.)

Ceiling panels should not be installed above plywood, composition board, or other insulating types of ceiling material. Surfaces of this type have an undesirable insulating effect that diminishes full heat output of the panel.

Radiant panels are often installed in floors. When this is done, the best arrangement is to place the coils in the concrete floor slab. Good results are obtained when the pipe or tubing is placed at least 2 inches below the floor surface, or deeper if a heavy traffic load is anticipated. Allow a minimum of two weeks for the concrete to set before applying heat, and then apply heat gradually. Floor covering of terrazzo, linoleum, tile, or carpeting can be installed. If the water temperature is kept below the prescribed maximum of 85°F, no damage will result to rugs, varnish, polish, or other materials. A typical piping diagram for a radiant floor panel is illustrated in Figure 15-13. Use care to avoid low places in the coil.

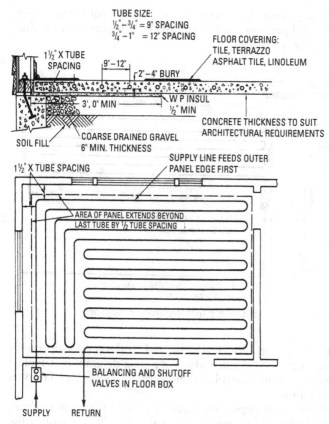

Figure 15-13 A typical radiant floor-panel diagram.

Heat loss to the ground from a floor panel laid directly on the ground can be expected. This loss is estimated to be from 10 to 20 percent of the heat provided for the room. Heat loss from outside slab edges can be greater than this percentage amount. Slab edges should receive from ½ inch to 2 inches of waterproof insulation.

It is not customary to install radiant wall panels except to provide supplementary heat where ceiling and/or floor panels do not provide a required degree of comfort. Wall panels are occasionally installed in bathrooms where more than normal heat temperatures may be desired. Figure 15-14 illustrates a typical radiant wall panel installation diagram. It must be remembered that circulating pumps for use with radiant heating should have a higher head rating than for convector systems of the same capacity. This requirement exists because the coil pressure drop is considerably higher than the drop in a radiator or convector.

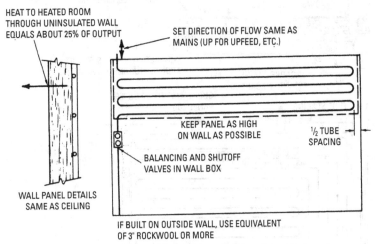

Figure 15-14 A radiant wall-panel installation diagram.

Certain fittings or devices are essential to the proper operation of a boiler. Figure 15-15 illustrates the piping diagram of a typical steam boiler with the control and indicating devices essential to its proper operation.

One of the most important features is the safety valve. Every boiler installation must have a safety valve installed to protect the boiler itself and the building occupants in case of malfunction. These valves are adjusted to open and relieve the internal pressure, should it rise above a safe predetermined level. Numerous valves, gages, and safety devices are found on all boilers.

The level of the water in a boiler must be maintained between certain limits; otherwise, serious damage to the boiler and building may result, as well as possible injury to the building occupants. Various safety devices are incorporated to protect against the possibility of this happening. Water gages are provided as a means of visually checking the level. More sophisticated gages employ floats that actuate a whistle or other alarm when the water level drops to a dangerous point.

Pressure-relief valves and fusible plugs are also used to protect the boiler in case of malfunction. Both of these devices will relieve dangerous high pressure under certain conditions.

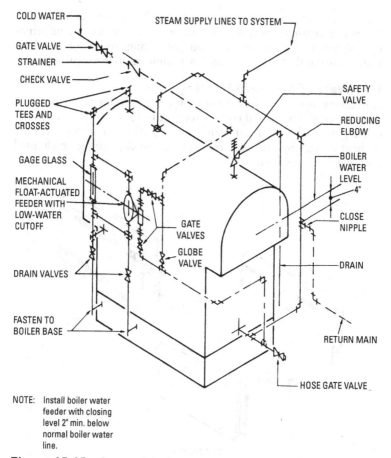

COLD WATER

STEAM SUPPLY LINES TO SYSTEM

GATE VALVE

STRAINER

CHECK VALVE

PLUGGED TEES AND CROSSES

SAFETY VALVE

REDUCING ELBOW

GAGE GLASS

MECHANICAL FLOAT-ACTUATED FEEDER WITH LOW-WATER CUTOFF

BOILER WATER LEVEL 4"

GATE VALVES

CLOSE NIPPLE

GLOBE VALVE

DRAIN VALVES

DRAIN

FASTEN TO BOILER BASE

RETURN MAIN

HOSE GATE VALVE

NOTE: Install boiler water feeder with closing level 2" min. below normal boiler water line.

Figure 15-15 A typical boiler showing the controls and indicating devices essential to its safe and proper operation.

Injectors are used to supply water to a boiler against the high pressure existing within. This is done by means of the jet principle. Steam loops are often provided to return condensate to a boiler. These devices are entirely automatic and have only one moving part—a check valve at the bottom of the drop leg.

Heating pumps are usually used in steam-heating systems to improve efficiency. Two principal types are available—condensation pumps and vacuum pumps. The type of heating system, its cost, and its individual requirements dictate which of these pumps must be used.

This heating section has provided a brief summary of steam and hot-water heating principles and, in general, contains the information necessary to answer questions pertaining to steam and hot-water heating that may be found in plumbing code examinations.

Hot-Water Baseboard Heating

A hydronic hot-water heating system circulates hot water to every room through baseboard panels (see Figure 15-16). *Hydronic* is another name for forced hot-water heating. The hot water gives off its heat energy by utilizing fins attached to the tubing or channel carrying the hot water.

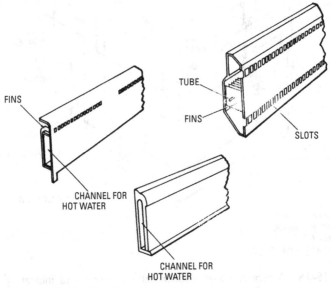

Figure 15-16 Baseboard panels.

Baseboard panels mounted around the outer perimeter of the home provide a curtain of warmth that surrounds those inside.

Radiant heat rays warm the room surfaces, but rising currents of convected warm air block out drafts and cold. Walls stay warm and cold spots are eliminated.

A boiler provides water between 120°F and 210°F. The water is pumped through the piping in the baseboard. Today's boilers are very efficient and very small in size. Most units are the size of an automatic washing machine, and some as small as suitcases and can be hung on the wall, depending on the size of the home being heated. Figure 15-17 shows an exploded view of a boiler.

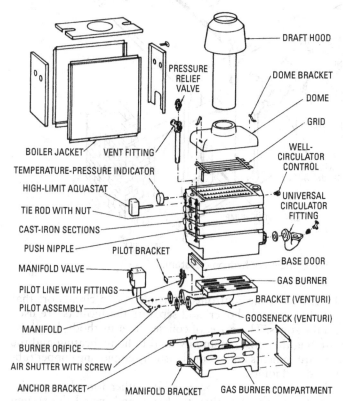

Figure 15-17 Exploded view of boiler for a hydronic system.

Circulating Pump
Booster pumps are used to circulate hot water through the pipes (see Figure 15-18). These pumps are designed to handle a wide range of pumping capacities. They will vary in size from small booster pumps with a 5-gallon-per-minute capacity to those capable of handling thousands of gallons per minute.

Piping Arrangements
There are two different piping arrangements utilized by the hydronic hot-water system. These are the *series loop* and the *one-pipe system*, which is utilized in zoning. The series-loop system is shown in

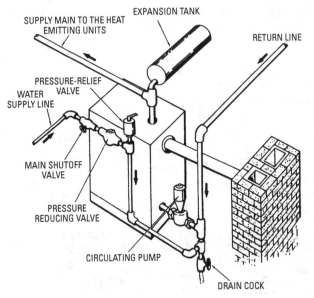

Figure 15-18 Parts of hydronic system.

Figure 15-19. The zone-controlled system has two circulators that are attached to a single boiler, and separate thermostats are used to control the zones. There are, of course, other methods, but these two are among the most commonly used in home heating and the plumbing requirements are minimal, usually employing copper tubing with soldered joints.

One-pipe systems may be operated on either forced or gravity circulation. Care must be taken to design and install the system with the characteristic temperature drop found in the heat emitting units farthest from the boiler in mind. Gravity circulation systems and the one-pipe system design must be planned very carefully to allow for heat load and losses in the system. The advantage of the one-pipe system lies in its ability to allow one or two heat-emitting units to be shut off without interfering with the flow of hot water to the other units. This is not the case in the series-loop system, where the units are connected in series and form a part of the supply line. Now examine Figure 15-19 and Figure 15-20 to see how the piping varies and how one allows the cut-off and the other does not. It is obvious that the series-loop system is less expensive to install, because it has a very simple piping arrangement.

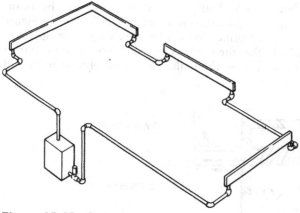

Figure 15-19 Series-loop system.

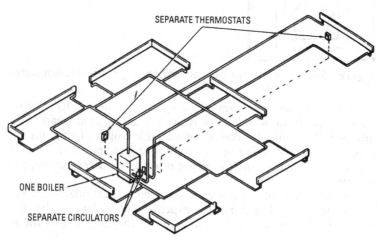

Figure 15-20 One-pipe, hot-water heating system with zoning.

Hot Water for Other Purposes

It is possible to use the boiler of a hydronic heating system to supply heat for such purposes as snow melting, heating a swimming pool, providing domestic hot water for household use, and other purposes. Separate circuits are created for each of these purposes, which are controlled by their own thermostats. Each is designed to tap into the main heating circuit from which it receives its supply of hot water.

For example, hot water for household use can be obtained by means of a heat exchanger or special coil inserted into the boiler (see Figure 15-21). The operating principles of a heat exchanger are shown in Figure 15-22. Note that the supply water does not come in contact with that being heated by the boiler for the baseboard units and other purposes.

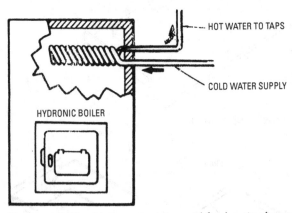

HOT WATER TO TAPS

COLD WATER SUPPLY

HYDRONIC BOILER

Figure 15-21   Heat exchanger used for heating household hot water.

One of the disadvantages of the hydronic system is its slow recovery time. If an outside door is opened during cold weather for any period of time, it takes a considerable length of time for the room to once again come up to a comfortable temperature. There is also the noise made from the piping heating up and expanding and popping.

The main advantage, however, is its economical operation. The type of fuel used determines the expense. The boilers can be electrically heated, heated with natural gas, or they may use oil for an energy source.

Word Completion Exercises

Complete the following sentences by filling in the blanks. (See Appendix C for answers.)

    1. Steam is an effective heating medium and is _____ to almost any type of building.

        a. adaptable

        b. useful

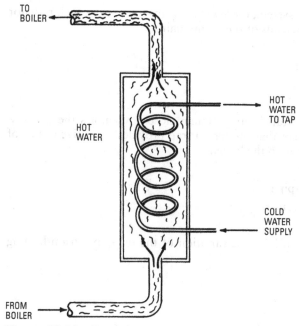

Figure 15-22 Principle of operation for a heat exchanger.

   c. useless

   d. dangerous

2. Large piping and radiator _____ are required to allow the condensate to return against the resistance offered by the steam flow.

   a. switches

   b. valves

   c. fins

   d. legs

3. The _____ gravity steam-heating system was developed to overcome the difficulty encountered when steam and condensate flow against one another in the same pipe.

   a. forced

   b. induced

   c. two-pipe

   d. single-pipe

4. In a ____ system, thermostatic traps are used at each radiator and at the ends of the steam mains.

  a. liquid
  b. heated
  c. hot
  d. vapor

5. A return-trap heating system closely resembles the ____ system, except that the return trap provides a positive return of condensate to the boiler.

  a. vapor
  b. single-pipe
  c. cold-water
  d. liquid

6. The ____ air-line is a variation of the one-pipe steam-heating system.

  a. filled
  b. vacuum
  c. wet
  d. dry

7. Piping and radiator valves must be oversized to accommodate the flow of both steam and condensate in the ____ heating system.

  a. large
  b. small
  c. vacuum-air-line
  d. gravity

8. The hot-water heating systems transmit only sensible heat to radiators, as distinguished from steam systems, which heat principally by the latent heat of _____.

  a. nature
  b. the sun
  c. condensation
  d. evaporation

9. The ____ hot-water systems are used mainly in small buildings such as homes and small business places.

  a. gravity

b. forced

c. liquid

d. cool

10. Radiant _____ heating is the method of heating a room by raising the temperature of one or more of its interior surfaces, such as floors, walls, or ceilings.

a. slab

b. panel

c. ceiling

d. floor

Chapter 16

License Requirements and Applications

Each state has its requirements for obtaining a license to be a journeyman plumber, a plumbing contractor, a master plumber, or an apprentice. Some states break licensing into specialty areas such as irrigation contractor, irrigation installer, appliance installer, mobile home contractor, mobile home installer, sewer and water installer, sewer and water contractor, water conditioning contractor, and water conditioning installer. The schedule of fees for each state varies so much that the amounts printed today may be changed tomorrow. Some states and local communities see these licensing fees as another way to support the plumbing department (both inspectors and administrators) without tapping into local tax revenues. Then again, other communities will accept the state plumber's license as proof of your qualifications.

Apprentices must serve a four-year apprenticeship or, in some instances (for specialized areas), a two-year apprenticeship is required. Documentation of experience is required by each state to become a journeyman, a master plumber, or a plumbing contractor. This is in addition to taking and passing the required exam for the area of specialization. Each exam has a healthy fee attached to it. All exams can be retaken, but this policy varies from state to state. There is a retake fee if you fail the first time.

Have the required background knowledge and have the required documents for certifying your experience and schooling properly notarized at the time you make application for the exam. The best way to obtain information is on the Internet. Each state has a Web site that details the requirements and lists the fees attached to each specialty area. Usually you can contact the proper office by www.<*name of the state*>.us.org

This chapter provides a sampling of states such as Alabama, Colorado, Delaware, Indiana, Michigan, Nevada, New Jersey, New Mexico, South Dakota, Texas, and Utah.

Some *cities,* such as Las Vegas, Nevada, require a license for the county/city in plumbing. New York City, of course, has its own licensing bureau with requirements and tests.

No matter where the test is taken, certain facts, figures, and techniques must be mastered before the state will grant a license and certify that the applicant is qualified to do the work covered by the license. This procedure is necessary to ensure that the water and sewage systems of the nation function properly and with maximum efficiency.

Alabama

Table 16-1 shows the requirements and fees for certification in the state of Alabama.

Table 16-1   Requirements and Fees for Certification

Certification	Requirements	Renewal Fees	Nonrefundable Exam Fees
Permanent Apprentice	No written exam required. Complete registration form.	$25.00	(One-time charge)
Journeyman Plumber	Requires two-year registration as apprentice or completed board-approved training program to take required written exam.	$30.00	$100.00
Journeyman Gas Fitter	Requires two-year registration as apprentice or completed board-approved training program to take required written exam.	$30.00	$100.00
Temporary-Revocable Journeyman Plumber or Gas Fitter	Requires approved application for written exam and expires 30 days past written exam date.	$15.00	(One-time charge)
Master Plumber	Requires one-year certification as journeyman plumber and required written exam.	$125.00	$150.00

Colorado

The State of Colorado application forms for journeyman plumber and master plumber are shown in Figures 16-1 and 16-2. Note the journeyman fee is $67 and the master plumber fee is $107. The fees are subject to change every July 1. Colorado has a state-wide plumbing license for residential, journeymen, and masters.

Colorado Division of Registrations
(303) 894-7800
www.dora.state.co.us/registrations

Reinstatement Application
JOURNEYMAN PLUMBER
Fee: $67.00

To reinstate your lapsed journeyman plumber license, you must:

- Complete and return this form to: Division of Registrations Office of Licensing, 1560 Broadway, Suite 1350, Denver, Colorado, 80202. For a name change, provide a copy of the legal document.
- Submit a reinstatement fee in the amount of **$67.00** either by check or money order payable to "State of Colorado" along with this form. **All fees are non-refundable and are subject to change every July 1.**

The content of this application must not be changed. If the content is changed, the applicant may be referred to the Colorado State Attorney General's Office for violation of Colorado law.

1. State of Colorado Journeyman Plumber's License No.: _____

2. Name: _____
   Last   First   Middle   Maiden

3. Mailing Address: _____
   Street & Apt. #   City   State   Zip Code

   This is a ☐ HOME   ☐ BUSINESS

4. Telephone: _____ E-mail Address: _____

5. Date of Birth: _____ U.S. Social Security Number:* _____

6. Date License Lapsed: _____

7. If your license has been lapsed for more than one year, please have a licensure verification sent to our office from all states where you have been licensed as a plumber. Please list on the line below all states where you hold, or ever held, a license to practice as a plumber. _____

8. Since the date your journeyman plumber license lapsed, have you been practicing as a journeyman plumber in the state of Colorado? _____ Yes _____ No

9. Are there any pending complaints against you in any other jurisdictions? _____ Yes _____ No

10. Have you ever had disciplinary action taken against you by another jurisdiction? _____ Yes _____ No

11. I state under penalty of perjury in the second degree, as defined in section 18-8-503, C.R.S., that the information contained in this application, to the best of my knowledge, is true and correct.

Signature of Applicant   Date

*Social Security Number Disclosure: Section 24-34-107(1) of the Colorado Revised Statutes requires that every application by an individual for a license issued pursuant to the authority set forth in title 12, C.R.S., by the Department of Regulatory Agencies, shall require the applicant's social security number. Disclosure of your social security number is mandatory for purposes of establishing, modifying, or enforcing child support under § 14-14-113 and § 26-13-126, C.R.S.; locating an individual who is under an obligation to pay child support as required by § 26-13-107(3)(a)(I)(A), C.R.S.; and reporting disciplinary actions to the Health Integrity and Protection Data Bank as required by 45 CFR §§ 61.1 *et seq.* Failure to provide your social security number for these mandatory purposes will result in the denial of your licensure application. Disclosure of your social security number is voluntary for disclosure to other state regulatory agencies, testing and examination vendors, law enforcement agencies, and other private federations and associations involved in professional regulation for identification purposes only. Your social security number will not be released for any other purpose unless provided for by law.

Updated 10/23/2003

Figure 16-1   Reinstatement application for journeyman plumber.

Colorado Division of Registrations
(303) 894-7800
www.dora.state.co.us/registrations

Reinstatement Application
MASTER PLUMBER
Fee $107.00

To reinstate your lapsed master plumber license, you must:

- Complete and return this form to: Division of Registrations Office of Licensing, 1560 Broadway, Suite 1350, Denver, Colorado, 80202. For a name change, provide a copy of the legal document.
- Submit a reinstatement fee in the amount of **$107.00** either by check or money order payable to "State of Colorado" along with this form. **All fees are non-refundable and are subject to change every July 1.**

The content of this application must not be changed. If the content is changed, the applicant may be referred to the Colorado State Attorney General's Office for violation of Colorado law.

1. State of Colorado Master Plumber's License No.: _____

2. Name: _____
   Last   First   Middle   Maiden

3. Mailing Address: _____
   Street & Apt. #   City   State   Zip Code

   This is a ☐ *HOME*   ☐ *BUSINESS*

4. Telephone: _____   E-mail Address: _____

5. Date of Birth: _____   U.S. Social Security Number:* _____

6. Date License Lapsed: _____

7. If your license has been lapsed for more than one year, please have a licensure verification sent to our office from all states where you have been licensed as a plumber. Please list on the line below all states where you hold, or ever held, a license to practice as a plumber. _____

8. Since the date your master plumber license lapsed, have you been practicing as a master plumber in the state of Colorado? _____ Yes _____ No

9. Are there any pending complaints against you in any other jurisdictions? _____ Yes _____ No

10. Have you ever had disciplinary action taken against you by another jurisdiction? _____ Yes _____ No

11. I state under penalty of perjury in the second degree, as defined in section 18-8-503, C.R.S., that the information contained in this application, to the best of my knowledge, is true and correct.

_____
Signature of Applicant   Date

*Social Security Number Disclosure: Section 24-34-107(1) of the Colorado Revised Statutes requires that every application by an individual for a license issued pursuant to the authority set forth in title 12, C.R.S., by the Department of Regulatory Agencies, shall require the applicant's social security number. Disclosure of your social security number is mandatory for purposes of establishing, modifying, or enforcing child support under § 14-14-113 and § 26-13-126, C.R.S.; locating an individual who is under an obligation to pay child support as required by § 26-13-107(3)(a)(I)(A), C.R.S.; and reporting disciplinary actions to the Health Integrity and Protection Data Bank as required by 45 CFR §§ 61.1 *et seq.* Failure to provide your social security number for these mandatory purposes will result in the denial of your licensure application. Disclosure of your social security number is voluntary for disclosure to other state regulatory agencies, testing and examination vendors, law enforcement agencies, and other private federations and associations involved in professional regulation for identification purposes only. Your social security number will not be released for any other purpose unless provided for by law.

Updated 10/23/2003

Figure 16-2 Reinstatement application for master plumber.

Delaware

The State of Delaware Board of Plumbing Examiners has statutory authority to grant plumbers licenses. There are five pages concerning the licensing of plumbers on the Internet (www.state.de.us/research/profreg/plubming.htm). As seen from Figure 16-3, the rules

1800 Board of Plumbing Examiners

PAGE < | TOP ▲ | TOC | PAGE > | PRINT

1800 Board of Plumbing Examiners

Statutory Authority: 24 **Del.C.** 1805(2)

1.0 General Provisions.

1.1 Legislative authority. These Rules and Regulations are adopted by the Delaware Board of Plumbing Examiners (hereinafter "the Board") by authority of 24 **Del.C.** Ch. 18, and the Administrative Procedures Act, 29 **Del.C.** Ch. 101.

1.2 Applicability. These Rules and Regulations shall govern proceedings before the Board to the extent they are consistent with governing law. Statutory reference: 29 **Del.C.** §10111 (2); 24 **Del.C.** §1805(2).

1.3 Officers. The Board will conduct election of officers for the offices of Chairperson, Vice-Chairperson and Secretary in May of each year. In the event of a resignation, termination or departure of one of the officers, a replacement shall be elected at the next Board meeting or at a meeting called for that purpose. Statutory reference: 24 **Del.C.** §1804(a).

1.4 Meetings. The Board shall, meet as often as necessary to transact the regular business of the Board and in any event, shall meet at least once each calendar quarter. Statutory reference: 24 **Del.C.** §1804(b).

1.5 Contact person. Information about the Board and its practices can be obtained by contacting the Division of Professional Regulation, Cannon Building, 861 Silver Lake Blvd., Ste. 203, Dover, Delaware 19904-2467, telephone 302-739- 4522 and http://www.professionallicensing.state.de.us/boards/plumbers/index.shtml. Statutory reference: 29 **Del.C.** § 10111(1).

2.0 Practice and Procedure.

2.1 Open meetings. All meetings of the Board, or any advisory or subcommittee, will be conducted in compliance with the Freedom of Information Act, 29 **Del.C.** Ch. 100. Statutory reference: 24 **Del.C.** §1812.

2.2 Disciplinary hearings. The procedural rules for disciplinary proceedings before the Board are outlined in 6.0. Statutory reference: 29 **Del.C.** §10111(2).

2.3 Advisory and subcommittees. The Board may appoint such advisory and subcommittees from time to time to assist in the performance of its duties as the Board deems necessary. Statutory reference. 29 **Del.C.** §10111(1).

3.0 Pre-examination Requirements for Licensure.

3.1 Definitions. The following definitions shall apply for purposes of this section:

3.1.1 "Performed plumbing services" means practical, hands-on experience working with tools in the installation, maintenance, extension, alteration, repair and removal of

http://www.state.de.us/research/profreg/plumbing.htm

Figure 16-3 Delaware Board of Plumbing Examiners page from Internet.

and regulations cover everything from general provisions to the voluntary treatment option for chemically dependent or impaired professionals.

Indiana

The State of Indiana has its application forms and answers to frequently asked questions (FAQs) posted on the Internet (www.in.gov/pla/bandc/plumbing/plumbing faq.html). Following are some of the Questions and Answers:

Question: How do you become an apprentice?

Answer: You must be registered in an approved program and be 17 years or older to qualify as an apprentice.

Question: How do you become a journeyman?

Answer: You must have at least 4 years in an apprenticeship program, or 4 years of experience in the plumbing field to qualify as a journeyman.

Question: How do you become a contractor?

Answer: You must have 4 years in an apprenticeship program or four years of experience in the plumbing trade or plumbing business under the direction of a licensed plumbing contractor.

Question: How many times can you take the test?

Answer: You can take the test seven times within a two-year period. However, if you do not pass the entire exam on first attempt, you shall be entitled to take the exam six additional times.

Question: Does reciprocity exist for journeyman and contractor plumbers?

Answer: Reciprocity is not available for journeyman and contractor plumbers. Instead, you must present an application and pay the appropriate fee. Then the application goes before the board. If the board approves, you must take the exam and pass.

Question: How much is the apprentice fee?

Answer: The apprentice fee is $10.00.

Question: How much is the journeyman fee?

Answer: The journeyman application fee is $15.00

Question: How much is the application fee for contractor?

Answer: The application fee for contractor is $30.00.

Question: May I work as a plumber without a license?

Answer: It is unlawful to act in the capacity of a plumbing contractor or journeyman plumber without a license.

The application form for approval of plumbing apprentice school displayed in Figure 16-4 shows the extent the state goes to make sure the apprenticeship programs are run on a professional level. The Indiana Professional Licensing Agency is responsible for making sure the schools meet the state laws and rules of the Indiana Plumbing Commission.

Note that in the application for journeyman plumber the examination fee is $30.00. The fees are nonrefundable and nontransferable in case you don't show up for the test.

Michigan

As in any state, a license is required to be a plumbing contractor. To qualify for the license, the applicant must hold a master plumber or employ the holder of a master plumber license as his or her representative. Only an owner of a sole proprietorship or partnership, or an officer of a corporation or limited liability company, may apply for licensure as a plumbing contractor.

The application for plumbing contractor shown in Figure 16-5 provides a glimpse of the requirements in this particular state. Other states are similar. In this case, the application is issued by the Michigan Department of Consumer & Industry Services, Bureau of Construction Codes & Fire Safety, and Plumbing Division in the state capital, Lansing.

Nevada (and Las Vegas)

The state of Nevada requires the passing of two exams, as well as a business and law exam. Las Vegas, Nevada, Clark County is an example of a larger city requiring licenses for both electricians and plumbers.

New Jersey

In New Jersey, the master plumber must complete five credits of continuing education every biennial renewal period. The State Board of Examiners of Master Plumbers is required to determine the topics required for each biennial renewal period and to put these topics in the New Jersey *Register.* In the 2003 to 2005 biennial renew period, the Board has decided to require every licensed master plumber to complete 1 hour of review of the existing National Standard Plumbing Code and one hour reviewing the laws and regulations pertinent to the practice of a master plumber (specifically, NJSA

Figure 16-4   Indiana application forms.

45:14C-1 et seq. and NJAC 13:32). The remaining 3 hours will be devoted to one or more of the following:

- Further review of the existing National Standard Plumbing Code
- Further review of laws and regulations pertinent to the practice of a master plumber

APPLICATION FOR JOURNEYMAN PLUMBER EXAMINATION FOR LICENSING
State Form 40602 (R10 / 11-02)
Approved by State Board of Accounts, 2002

Indiana Professional Licensing Agency
302 W. Washington St., Rm. E034
Indianapolis, IN 46204-2700
(317)-232-2980
www.in.gov/pla

FEE: $30.00   ALL FEES ARE NON-REFUNDABLE AND NON-TRANSFERABLE.

Social Security number *

* Your Social Security number is requested by this agency in accordance with IC 4-1-8-1; it is mandatory that it be given. Social Security numbers are made available to the Department of Revenue.

Name of applicant

Date of birth (month, day, year)

Address (number and street, city, state, ZIP code)

County

Telephone number

Have you ever been convicted of a crime? (if "Yes", provide a copy of the court order and any pertinent documents)
☐ Yes   ☐ No

INSTRUCTIONS:
1. If you are applying on the basis of having completed four (4) years of training in an approved apprenticeship program, please complete Sections one (1) and three (3).
2. If you are applying on the basis of having completed four (4) years of experience in the plumbing trade, please complete Sections two (2) and three (3).

SECTION ONE

APPROVED APPRENTICESHIP PROGRAM SPONSOR CERTIFICATION OF COMPLETION

I have successfully completed the following four (4) years of training in an approved apprenticeship program, satisfying the requirements as defined in commission rule, 860 IAC 1-1-9, as verified by the sponsor of the approved apprenticeship program, herein:

Name of apprenticeship program sponsor

Telephone number

Address (number and street, city, state, ZIP code, county)

Date of enrollment (month, year)

Date of completion (month, year)

I hereby certify that _____ successfully
                                    Name of apprentice
completed four (4) years of training in an approved apprenticeship program.

Date of enrollment

Signature of manager of approved apprenticeship program sponsor

Date of completion

Date signed

NOTARY CERTIFICATE

STATE OF _____

COUNTY OF _____   } SS:

I, _____, having been duly sworn on oath, say that I am the
above-named, that I have personally prepared the foregoing affidavit, and that the same is true to the best of my knowledge and belief.

Signature of manager of approved apprenticeship program sponsor

Signature of Notary Public

Printed or typed name of manager of approved apprenticeship program sponsor

Printed or typed name of Notary Public

Date subscribed and sworn to Notary Public

County of residence

Date commission expires

Figure 16-4a (Continued)

- Job site safety, including the Occupational Safety and Health Administration (OSHA) and related information
- Americans with Disabilities Act (ADA) barrier-free provisions and Uniform Construction Code (UCC) requirements
- Plumbing systems pipe sizing for waste, drainage, vent, water, and gas
- Contract liability, insurance requirements, business law, consumer contracts, and ethical conduct

I have completed the following four (4) years of experience in the plumbing trade, satisfying the requirements as defined in commission rule, 860 IAC 1-1-9 and 860 IAC 1-1-10, as verified by employer, attached herewith:

Name of employer	Plumbing contractor license number (if applicable): PC
Address (number and street, city, state, ZIP code)	
County	Telephone number
Dates of employment (month, day, year): From	To
Name of employer	Plumbing contractor license number (if applicable): PC
Address (number and street, city, state, ZIP code)	
County	Telephone number
Dates of employment (month, day, year): From	To

APPLICANT AFFIDAVIT OF EXPERIENCE IN PLUMBING TRADE

I hereby certify that I, _____ have worked in the plumbing trade as defined in commission rule 860 IAC 1-1-9, for
_____ Name of applicant

the period of _____ to _____, for _____.
Day, month, year   Day, month, year   Name of company or plumbing business

Name of employer or licensed contractor

Address (number and street, city, state, ZIP code)

I further certify that I am unable to obtain an employer affidavit verifying the aforementioned experience in the plumbing trade due to the following reason(s):

Signature of applicant	Date signed

NOTARY CERTIFICATE

STATE OF _____
COUNTY OF _____ } SS:

I, _____, having been duly sworn on oath, say that I am the above-named, that I have personally prepared the foregoing affidavit, and that the same is true to the best of my knowledge and belief.

Signature of applicant	Signature of Notary Public	
Printed or typed name of applicant	Printed or typed name of Notary Public	
Date subscribed and sworn to Notary Public	County of residence	Date commission expires

Figure 16-4b (Continued)

- The licensed master plumber's responsibilities for an employee's workmanship and competence
- International Mechanical Code requirements
- Backflow prevention and backflow devices
- New Jersey sales tax and other tax matters
- Organizing/managing a business, estimating/bidding, and project managing
- Excavation and backfill requirements and procedures

Figure 16-4c (Continued)

New Mexico

In New Mexico, the Contractor's Licensing Service, Inc. in Albuquerque provides a free newsletter, services, books, and programs for obtaining the plumber's license. These programs (as well as official programs in technical schools and two-year colleges) are available from the Plumbing License Bureau or Division in each state. Licensing comes under the Attorney General's office in some states, the State Plumbing Board or State's Plumbing Bureau or the State Plumbing Commission. The Internet is a good place to look for the proper licensing authority in any state. Contractor, master, journeyman, and apprentice applications are available from these licensure boards. They also have a schedule of when and where the examinations are given and the fee for each. It is best to know what class of license is being sought at the time of application.

New Mexico has many journeyman and contractor license classifications. New Mexico requires the passing of two exams, as well as a business and law exam.

Application For Plumbing Contractor License (Grandfathered)
Michigan Department of Consumer & Industry Services
Bureau of Construction Codes & Fire Safety
Plumbing Division
P.O. Box 30254
Lansing, MI 48909
517/241-9330

No Fee Required

Authority: 2002 PA 733 Completion: Mandatory Penalty: License will not be issued	The Department of Consumer and Industry Services will not discriminate against any individual or group because of race, sex, religion, age, national origin, color, marital status, disability or political beliefs. If you need help with reading, writing, hearing, etc., under the Americans with Disabilities Act, you may make your needs known to this agency.

Instructions
•A person applying for licensure as a Plumbing Contractor must complete this application and submit all required information to the address listed above.

The provisions of 2002 PA 733 states:

To qualify for a plumbing contractor license, the applicant must either hold a master plumber license or employ the holder of a master plumber license as his or her representative. Only an owner of a sole proprietorship or partnership, or an officer of a corporation or limited liability company, may apply for licensure as a plumbing contractor.

A person who, on the effective date of this act, is licensed as a master plumber under former 1929 PA 266 or employing a licensed master plumber shall, upon payment of the plumbing contractor license fee and upon furnishing the department with satisfactory evidence of having been engaged in a business as a master plumber for a minimum of 3 out of the 5 years immediately preceding the effective date of this act, be granted a plumbing contractor license without examination if the person applies within 6 months after the effective date of this act.

An individual licensed under this act employed or acting as a plumbing inspector shall not engage in, or be directly or indirectly connected with, the plumbing business including, but not limited to, the furnishing of labor, materials, or appliances for the construction, alteration, or maintenance of a building or the preparation of plans or specifications for the construction, alteration, or maintenance of a building and shall not engage in any work that conflicts with his or her official duties.

• **You must return your current master plumber (wall and pocket) license with this application. Retain a copy of this application and your license until your new license is received.**

• Plumbing contractors shall provide one of the following:
•A notarized letter stating you are the sole proprietor.
•Copies of partnership papers.
•Copies of incorporation papers.

• Social Security Number. A person may be exempt from providing this information under 1996 PA 236. A person is not required to include this information when exempt under this act from obtaining a social security number or for religious convictions prohibiting the disclosure of this information.

Applicant Information

NAME (Last, Name, First Name, Middle Name)	DATE OF BIRTH	SOCIAL SECURITY NUMBER	HOME TELEPHONE NUMBER
HOME ADDRESS	CITY	STATE	ZIP CODE
BUSINESS NAME	TYPE OF BUSINESS	DATES OF EMPLOYMENT (MM/DD/YY) FROM: TO:	
BUSINESS ADDRESS	COUNTY	TOWNSHIP	
CITY	STATE	ZIP CODE	BUSINESS TELEPHONE NUMBER
NAME OF BUSINESS OWNER OR PRESIDENT OF CORPORATION, IF DIFFERENT THAN APPLICANT		TITLE	
NAME OF MASTER PLUMBER		LICENSE NUMBER	
HOME ADDRESS	CITY	STATE	ZIP CODE

BCC-945 (4/03) Front

Figure 16-5 Michigan application forms.

South Dakota

The State of South Dakota's State Plumbing Commission has as its mission statement the following:

Previous Business Information - In accordance with law, applicants must provide complete business information for the previous 5 years. If the information contained under "Applicant Information" does not cover the 5-year period, please complete the information requested below. (Attach additional sheets if necessary)

PREVIOUS BUSINESS NAME			TYPE OF BUSINESS		
PREVIOUS BUSINESS ADDRESS				DATES OF OPERATION (MM/DD/YY) FROM: TO:	
CITY	STATE	ZIP CODE	TOWNSHIP	COUNTY	
NAME OF BUSINESS OWNER OR PRESIDENT OF CORPORATION				TITLE	
PREVIOUS BUSINESS NAME			TYPE OF BUSINESS		
PREVIOUS BUSINESS ADDRESS				DATES OF OPERATION (MM/DD/YY) FROM: TO:	
CITY	STATE	ZIP CODE	TOWNSHIP	COUNTY	
NAME OF BUSINESS OWNER OR PRESIDENT OF CORPORATION				TITLE	

Branch Information - Provide the information below for each branch office operated by your company. (Attach additional sheets if necessary.)

A licensed plumbing contractor may operate 1 or more branch offices in this state bearing the same firm name provided a licensed master plumber is in charge and has the responsibility of supervision at each branch.

NAME OF MASTER PLUMBER					LICENSE NUMBER
BRANCH ADDRESS	CITY		STATE	ZIP CODE	BUSINESS TELEPHONE NUMBER
NAME OF MASTER PLUMBER					LICENSE NUMBER
BRANCH ADDRESS	CITY		STATE	ZIP CODE	BUSINESS TELEPHONE NUMBER
NAME OF MASTER PLUMBER					LICENSE NUMBER
BRANCH ADDRESS	CITY		STATE	ZIP CODE	BUSINESS TELEPHONE NUMBER

Background Information

Have you been convicted of a felony or misdemeanor?

☐ No   ☐ Yes

If yes, you will be provided with a "Request for Conviction History" form after filing this application. Failure to accurately respond to this question will result in you forfeiting any rights of consideration for examination and issuance of a plumber's license in the state of Michigan.

If the master plumber listed above for branch locations is not the person seeking licensure as the contractor, the applicant must provide an original notarized letter for each master plumber stating the master plumber at that branch will be in the full time employment of the contractor and will be actively in charge of and responsible for code compliance of all installation of plumbing must accompany this application.

Certification and Signature

I certify that the information provided is true and accurate to the best of my ability. I further understand that falsification of any statement is cause for rejection of application or revocation of license, if issued.

BUSINESS OWNER'S NAME (TYPE OR PRINT)	
BUSINESS OWNER'S SIGNATURE	DATE

BCC-945 (4/03) Back

Figure 16-5a   (Continued)

To protect the public from unsafe drinking water and unsafe waste disposal facilities by licensing qualified plumbers; to inspect plumbing installations and ensure that the state plumbing code is updated and distributed; to inform plumbers, inspection departments and the public of code requirements, new products and methods of installation; and, to utilize seminars and the media to provide information of the board's activities, recommendations and requirements.

The South Dakota State Plumbing Commission is responsible for granting plumbing licenses to qualified candidates. The State Plumbing Commission administers and establishes educational, training, and examination requirements for plumbing contractor, journeyman, and apprentice; sewer and water contractor, installer, and apprentice; appliance contractor, installer, and apprentice; water conditioning contractor, installer, and apprentice; mobile home contractor, installer, and apprentice; and underground irrigation contractor, installer, and apprentice. Plumbing apprentices must serve a 4-year apprenticeship. All other apprentices must serve a 2-year apprenticeship. Journeyman plumbers must work for two years before becoming a plumbing contractor. All installers must work for 1 year to become a contractor.

Exams are held in Pierre at any time by advance appointment. Exams are also held in Sioux Falls and Rapid City but not on specific preset dates. Exams are held in these locations when the office has received enough applicants to warrant travel to those sites. Eight to 10 applicants are minimum to warrant travel to those sites.

Fees

A schedule of fees for first-time exams are listed in Table 16-2. All retakes of the exam are $50.00.

Table 16-2 Schedule of Fees in South Dakota

Exam	Fee
Appliance installation contractor	$175
Appliance installer	$100
Irrigation contractor	$175
Irrigation installer	$100
Journeyman plumber	$130
Mobile home contractor	$175
Mobile home installer	$100
Plumbing contractor	$260
Sewer and water contractor	$225
Sewer and water installer	$100
Temporary permit	$50
Water conditioning contractor	$175
Water conditioning installer	$100

State Requirements

Requirements for becoming each of these listed installers, contractors, and plumbers are given in the South Dakota State Plumbing Commission's Web site (www.state.sd.us/dol/boards/plumbing/licens.htm). For those without access to the Internet, the following sections are presented with the commission's permission.

Becoming a Plumbing Contractor/Journeyman Plumber/Apprentice

Plumbing contractors, journeyman plumbers, and apprentice plumbers may engage in the furnishing and/or the use of materials and fixtures in the installation, extension, and alteration of all piping, fixtures, appliances, and appurtenances in connection with sanitary drainage or storm drainage facilities, the venting system and the public water supply systems, within or adjacent to any building, structure, or conveyance; and including the installation, extension, or alteration of the storm water, liquid waste, or sewerage and water supply systems of any premises to their connection with any point of public disposal, or other regulated terminal.

- Applicants for a plumbing contractor license shall show evidence of 6 years experience as a plumbing contractor, plumber, or plumber's apprentice with at least two of those years as a plumbing contractor or plumber.

- During the 6 years, the applicant must have spent at least 1900 hours per year as a plumbing contractor, plumber, or plumber's apprentice.

- Applicants for a plumbing contractor's license must fill out an application and pay an examination and license fee of $260.

- Applicants for a journeyman plumber's license shall show evidence of 4 years experience as an apprentice plumber.

- During the 4 years, the applicant must have spent at least 1900 hours per year as an apprentice.

- Credit for military plumbing is given at the rate of 1 year credit for each 2 years in the military, up to a maximum of 5 years of credit.

- Applicants for a journeyman plumber's license must fill out an application and pay an examination and license fee of $130.

- If the commission finds the individuals have the required experience, they shall be tested.

- Applicants for a plumber apprentice license must fill out an application showing the plumber under which the applicant is working.

- The license is without charge for the first 4 years and then the fee is $80.
- Apprentice plumbers who have had 3 years (5700 hours) experience in learning and assisting in the installation, alteration, and repair of plumbing under a plumbing contractor may work during their 4th year of apprenticeship without being under the direct supervision of a plumbing contractor or plumber.

Becoming a Sewer and Water Contractor/Installer/Apprentice

Sewer and water contractors, installers, and apprentices may engage in the setting up of building sewer and water service, the repair of existing building sewer and water services, the setting up of building storm sewer, the repair of existing building storm sewer, and the setting up of water treatment plant piping and equipment designed to purify water, chemical treatment piping, and sewer treatment plant piping and equipment designed to treat sewage, and the repair of the piping and equipment.

- Applicants for a sewer and water contractor shall show evidence of 1 year of experience as a sewer and water installer.
- Applicants for a sewer and water installer shall show evidence of 2 years experience as a sewer and water apprentice.
- If the commission finds the individuals have the required experience, they shall be tested.
- Applicants must fill out an application form.
- A sewer and water contractor shall pay an examination and license fee of $225.
- A sewer and water installer shall pay an examination and license fee of $100.
- All applicants for a sewer and water apprentice license shall fill out an application showing the sewer and water installer under which the applicant is working. The license fee is without charge for 2 years, and then the fee is $50.

Becoming an Appliance Contractor/Installer/Apprentice

Appliance contractors, installers, and apprentices may engage in the making of local connections to water and waste systems of all miscellaneous and commercial equipment designed to use electricity or gas (or both) as a basic source of energy; maintenance and service of this equipment, including operation, adjustment, repair, removal, and renovation; but not connection and repair of household and commercial plumbing fixtures that are designed for sanitary use or water conditioning.

- Applicants for an appliance contractor license shall show evidence of 1 year of experience as an appliance installer.
- Applicants for an appliance installer license shall show evidence of 2 years experience as an appliance apprentice.
- If the commission finds the individuals have the required experience, they shall be tested.
- Applicants must fill out an application form.
- An appliance contractor shall pay an examination and license fee of $175.
- An appliance installer shall pay an examination and license fee of $100.
- All applicants for an appliance apprentice shall fill out an application showing the appliance installer under which the applicant is working. The license is without charge for 2 years and then the fee is $50.

Becoming a Water Conditioning Contractor/Installer/Apprentice

Water conditioning contractors, installers, and apprentices may engage in the treatment of water and the installation of appliances, appurtenances, fixtures, and plumbing necessary thereto, all designed to treat water to alter, modify, add, or remove mineral, chemical, or bacterial content, and to repair such equipment to a water distribution system. Water conditioning installation, repair, and treatment do not mean the exchange of appliances, appurtenances, and fixtures when the plumbing has previously been installed or adapted for such appliances, appurtenances, and fixtures, and no substantial change in the plumbing system is required.

- Applicants for a water conditioning contractor shall show evidence of one year of experience as a water conditioning installer.
- Applicants for a water conditioning installer shall show evidence of 2 years experience as a water conditioning apprentice.
- If the commission finds the individuals have the required experience, they shall be tested.
- Applicants must fill out an application form.
- A water conditioning contractor shall pay an examination and license fee of $175.

- A water conditioning installer shall pay an examination and license fee of $100.
- All applicants for a water conditioning apprentice shall fill out an application showing the water conditioning installer under which the applicant is working. The license fee is without charge for 2 years and then the fee is $50.

Becoming a Mobile Home Contractor/Installer/Apprentice

Mobile home contractors, installers, and apprentices may engage in the connection to local water and waste systems from manufactured and mobile homes only, including maintenance and service (which include the operation, adjustment, repair, removal and renovation of such connections).

- Applicants for a mobile home contractor license shall show evidence of 1 year of experience as a mobile home installer.
- Applicants for a mobile home installer license shall show evidence of 2 years experience as a mobile home apprentice.
- If the commission finds the individuals have the required experience, they shall be tested.
- Applicants must fill out an application form.
- A mobile home contractor shall pay an examination and license fee of $175.
- A mobile home installer shall pay an examination and license fee of $100.
- All applicants for a mobile home apprentice shall fill out an application showing the mobile home installer under which the applicant is working. The license is without charge for two years and then the fee is $50.

Becoming an Underground Irrigation Contractor/Installer/Apprentice

Underground irrigation contractors, installers, and apprentices may engage in the practice of and the furnishing or the use of materials and devices for the purpose of installing underground irrigation systems and connecting them to the local source of potable water, including maintenance and service (which includes the operation, adjustment, repair, removal, and renovation of such connections and devices).

- Applicants for an underground irrigation contractor shall show evidence of 1 year of experience as an underground irrigation installer.

SOUTH DAKOTA STATE PLUMBING COMMISSION
118 West Capitol
Pierre, South Dakota 57501-2017
Phone: (605) 773-3429
FAX: (605) 773-5405

INSTRUCTIONS FOR COMPLETING THIS APPLICATION

1. This application must be typewritten or printed in ink.
2. Complete all spaces provided. If the Question does not apply, write "none" in the space.
3. Accompany this application with the appropriate fee as indicated on the application.
4. Accompany this application with written statements from present and previous employers which must state: a) dates of employment; b) number of hours worked during employment; and, c) extent of work performed during employment.
5. Reciprocity applicants - accompany this application with a photo-copy of your current license from the state in which you are licensed and disregard instruction #5.
(MN, ND, MT, CO RESIDENTS ONLY)
6. Applicants for apprentice licenses may disregard instructions #5 and #6.

NOTE: FAILURE TO COMPLY WITH ALL INSTRUCTIONS WILL CAUSE APPLICATION TO BE RETURNED!

- PLUMBING CONTRACTOR.................$_____
- JOURNEYMAN PLUMBER.....................$_____
- APPRENTICE PLUMBER............................FREE
  (Less than 4 years)
- APPRENTICE PLUMBER (Fifth year).......$_____
- APPLIANCE INSTALLATION
  CONTRACTOR............................$_____
- APPLIANCE INSTALLATION
  INSTALLER...............................$_____
- APPL. INST. APPRENTICE.............................FREE
  (Less than 2 years)
- APPL. INST. APPRENTICE (Third year).....$_____
- MOBILE HOME CONTRACTOR..............$_____
- MOBILE HOME INSTALLER.................$_____
- MOBILE HOME APPRENTICE................$_____
- TEMPORARY PERMIT........................$_____

- WATER CONDITIONING CONTRACTOR...$_____
- WATER CONDITIONING INSTALLER........$_____
- W/C INST. APPRENTICE................................FREE
  (Less than 2 years)
- W/C INST. APPRENTICE.....................$_____
  (Third year)....................................$_____
- SEWER & WATER CONTRACTOR..........$_____
- SEWER & WATER INSTALLER...............$_____
- S. & W. INST. APPRENTICE............................FREE
  (Less than 2 years)
- S. & W. INST. APPRENTICE...................$_____
  (Third year)....................................$_____
- IRRIGATION CONTRACTOR..................$_____
- IRRIGATION INSTALLER......................$_____
- IRRIGATION APPRENTICE...................$_____

NAME_____ DATE_____
SOCIAL SECURITY NUMBER (_____) RES. PHONE NUMBER_____
RESIDENCE ADDRESS_____
Street Number   County   City   State   Zip Code
YOUR AGE____ YOUR DATE OF BIRTH____ YOUR PLACE OF BIRTH____
Mo. Date Year   City   State
PRESENT EMPLOYER_____ WORK PHONE NUMBER_____
EMPLOYED AS_____
ADDRESS OF EMPLOYER_____
PB201B Rev: 8-93-500

NEED BOARD SIGNING
SEND REFERENCES (_____)

Figure 16-6 South Dakota application forms.

- Applicants for an underground irrigation installer shall show evidence of 2 years experience as an underground irrigation apprentice.
- If the commission finds the individuals have the required experience, they shall be tested.
- Applicants must fill out an application form.
- An underground irrigation contractor shall pay an examination and license fee of $175.

Have you ever carried a Plumbing License?_____ If so, where?_____

State the type or grade of License_____ In force, from _____ to _____

Was the License obtained by examination?_____ Have you ever had a Plumbing License revoked? _____ By whom?_____

If so, give reasons_____

Have you previously been examined for a Plumbing License by this commission?_____

If so, state type, and results of examination _____

Approved ?

Disapproved ?

Have you previously made application for a State of South Dakota Plumbing License?_____

SCHOOL RECORD

Education: Circle Highest Grade Completed   1   2   3   4   5   6   7   8   9   10   11   12   13   14   15   16

Are you a graduate of a Plumbing Course of an accredited University or College?_____

Give degree_____ Year_____ Name of School_____

Address of School _____

Are you a graduate of a Plumbing Trade School?_____

Name of above School_____

Address of above School_____

State other courses of Plumbing Study, if any_____

Name and address of above_____

EMPLOYMENT DATA

Be sure that you break down your experience according to each classification.

Total years of Plumbing Experience	EXPERIENCE					
	As Apprentice		As Journeyman		As Contractor	
CLASSIFICATION	Months	Years	Months	Years	Months	Years
Residential Plumbing.....................						
Commercial & Industrial Plumbing....						
Farmstead Plumbing.....................						
Plumbing Maintenance & Repair......						
Sewer & Water Installation.............						
Appliance Installation...................						
Water Cond't. Installation..............						
Planning & laying out for..............						
Mobile Home Plumbing Work.........						
	TOTAL YEARS		TOTAL YEARS		TOTAL YEARS	

Figure 16-6a (Continued)

- An underground irrigation installer shall pay an examination and license fee of $100.
- All applicants for an underground irrigation apprentice shall fill out an application showing the underground irrigation installer under which the applicant is working. The license is without charge for 2 years and then the fee is $50.

REFERENCES

List at least three (3) persons actively engaged in the plumbing industry that you have worked under.

Name_____	Name_____
Address_____	Address_____
Occupation_____	Occupation_____
Name_____	Name_____
Address_____	Address_____
Occupation_____	Occupation_____

PLUMBING EMPLOYMENT RECORD

IMPORTANT Unless complete address of employer is given, it is impossible to properly process your application and will cause delay. **PREVIOUS AND PRESENT EMPLOYERS**	DATES EMPLOYED From Month Year	To Month Year	TYPE OF PLUMBING WORK
Name Address		Present	
	From	To	
Name Address			
	From	To	
Name Address			

I declare and affirm under the penalties of perjury that this claim (petition, application, information) has been examined by me, and to the best of my knowledge and belief, is in all things true and correct.

_____
SIGNATURE

Figure 16-6b (Continued)

Application Form

The South Dakota application form for the plumber, contractor, installer, and apprentice is shown in Figure 16.6 for easy reference and guidance purposes. It should give you an idea as to what is needed to apply for a license or apprenticeship.

Texas

In Texas, the license is for journeyman, master, and, if an electrician, it can also be for signs. Air-conditioning licenses in Texas may be

REMARKS BY APPLICANT

_____

_____

_____

_____

_____

_____

_____

_____

_____

_____

_____

SPACE BELOW RESERVED FOR COMMISSION

Approved_____ Disapproved_____ for examination. Types of examination_____ Date_____

By _____ and _____
    Contractor Member   Plumber Member

RECORD	Verified	Not Verified
Training		
Work Experience		
Work Reference		

Corrected by_____

Checked by_____

License #_____ as _____

#1	
#2	
#3	
#4	
#5	

Examination   #1   2   3   4   5   Date of Examination_____ Exam Supervised By_____

RATE AS _____ A

By_____ Date_____

DATE_____
GRADE_____
PASSED   FAILED
BY_____

Notes: _____
_____
_____
_____
_____

LICENSE NUMBER ISSUED_____
DATE_____ INITIALS_____

Figure 16-6c (Continued)

divided into A and B. The A license is unlimited tonnage of air conditioning, while the B license is limited to work with 15 tons or less.

Utah

In Utah, the statewide plumbing license is for residential, journeyman, and master. Utah requires the passing of two exams, as well as a business and law exam.

Appendix A

Miscellaneous Information

This appendix contains useful reference information, including the following:

- Decimal and millimeter equivalents of fractional parts of an inch
- Commercial pipe sizes and wall thicknesses
- American standard graphical symbols for pipe fittings and valves
- American standard graphical symbols for piping
- Weights and measures

Commercial Pipe Sizes and Wall Thicknesses

The subsequent table lists the pipe sizes and wall thicknesses currently established as standard, or specifically the following:

1. The traditional standard weight, extra strong, and double extra strong pipe.
2. The pipe wall thickness schedules listed in American Standard B36.10, which are applicable to carbon steel and alloys other than stainless steels.
3. The pipe wall thickness schedules listed in American Standard B36.19, which are applicable only to stainless steels.

Decimal and Millimeter Equivalents of Fractional Parts of an Inch

Parts of Inch	Decimal	Millimeters	Parts of Inch	Decimal	Millimeters
$1/64$	0.01563	0.397	$33/64$	0.51563	13.097
$1/32$	0.03125	0.794	$17/32$	0.53125	13.097
$3/64$	0.04688	1.191	$35/64$	0.54688	13.890
$1/16$	0.0625	1.587	$9/16$	0.5625	14.287
$5/64$	0.07813	1.984	$37/64$	0.57813	14.684
$3/32$	0.09375	2.381	$19/32$	0.59375	15.081
$7/64$	0.10938	2.778	$39/64$	0.60938	15.478
$1/8$	0.125	3.175	$5/8$	0.625	15.875
$9/64$	0.14063	3.572	$41/64$	0.64063	16.272
$3/32$	0.15625	3.969	$21/32$	0.65625	16.669
$11/64$	0.17188	4.366	$43/64$	0.67188	17.065
$3/16$	0.1875	4.762	$11/16$	0.6875	17.462
$13/64$	0.20313	5.159	$45/64$	0.70313	17.859
$7/32$	0.21875	5.556	$23/32$	0.71875	18.256
$15/64$	0.23438	5.953	$47/64$	0.73438	18.653
$1/4$	0.25	6.350	$3/4$	0.75	19.050
$17/64$	0.26563	6.747	$49/64$	0.76563	19.447
$9/32$	0.28125	7.144	$25/32$	0.78125	19.844
$19/64$	0.29688	7.541	$51/64$	0.79688	20.240
$5/16$	0.3125	7.937	$13/16$	0.8125	20.637
$21/64$	0.32813	8.334	$53/64$	0.82813	21.034
$11/32$	0.34375	8.731	$27/32$	0.84375	21.431
$23/64$	0.35938	9.128	$55/64$	0.85938	21.828
$3/8$	0.375	9.525	$7/8$	0.875	22.225
$25/64$	0.39063	9.922	$57/64$	0.89063	22.622
$13/32$	0.40625	10.319	$29/32$	0.90625	23.019
$27/64$	0.42188	10.716	$59/64$	0.92188	23.415
$7/16$	0.4375	11.113	$15/16$	0.9375	23.812
$29/64$	0.45313	11.509	$61/64$	0.95313	24.209
$15/32$	0.46875	11.906	$31/32$	0.96875	24.606
$31/64$	0.48438	12.303	$63/64$	0.98438	25.003
$1/2$	0.5	12.700	1	1.00000	25.400

Nominal Wall Thickness For

Nominal Pipe Size	Outside Diam.	Sched. 5*	Sched. 10*	Sched. 20	Sched. 30	Standard†	Sched. 40	Sched. 60	Extra Strong‡	Sched. 80	Sched. 100	Sched. 120	Sched. 140	Sched. 160	XX Strong
1/8	0.405	–	0.049	–	–	0.068	0.068	–	0.095	0.095	–	–	–	–	–
1/4	0.540	–	0.065	–	–	0.068	0.086	–	0.119	0.119	–	–	–	–	–
3/8	0.675	–	0.065	–	–	0.091	0.091	–	0.126	0.126	–	–	–	–	–
1/2	0.840	–	0.083	–	–	0.109	0.109	–	0.147	0.147	–	–	–	0.187	0.294
3/4	1.050	0.065	0.083	–	–	0.113	0.113	–	0.154	0.154	–	–	–	0.218	0.308
1	1.315	0.065	0.109	–	–	0.133	0.133	–	0.179	0.179	–	–	–	0.250	0.358
1¼	1.660	0.065	0.109	–	–	0.140	0.140	–	0.191	0.191	–	–	–	0.250	0.382
1½	1.900	0.065	0.109	–	–	0.145	0.145	–	0.200	0.200	–	–	–	0.281	0.400
2	2.375	0.065	0.109	–	–	0.154	0.154	–	0.218	0.218	–	–	–	0.343	0.436
2½	2.875	0.083	0.120	–	–	0.203	0.203	–	0.276	0.276	–	–	–	0.375	0.552
3	3.5	0.082	0.120	–	–	0.216	0.216	–	0.300	0.300	–	–	–	0.438	0.600
3½	4.0	0.083	0.120	–	–	0.226	0.226	–	0.318	0.318	–	–	–	–	–
4	4.5	0.083	0.120	–	–	0.237	0.237	–	0.337	0.337	–	0.438	–	0.531	0.674
5	5.563	0.109	0.134	–	–	0.258	0.258	–	0.375	0.375	–	0.500	–	0.625	0.750
6	6.625	0.109	0.134	–	–	0.280	0.280	–	0.432	0.432	–	0.562	–	0.718	0.864
8	8.625	0.109	0.148	0.250	0.277	0.322	0.322	0.406	0.500	0.500	0.593	0.718	0.812	0.906	0.875
10	10.75	0.134	0.165	0.250	0.307	0.365	0.365	0.500	0.500	0.593	0.713	0.843	1.000	1.125	–
12	12.75	0.156	0.180	0.250	0.330	0.375	0.406	0.562	0.500	0.687	0.843	1.000	1.125	1.312	–
14 O.D.	14.0	–	0.250	0.312	0.375	0.375	0.438	0.593	0.500	0.750	0.937	1.093	1.250	1.406	–
16 O.D.	16.0	–	0.250	0.312	0.375	0.375	0.500	0.656	0.500	0.843	1.031	1.218	1.438	1.593	–
18 O.D.	18.0	–	0.250	0.312	0.438	0.375	0.562	0.750	0.500	0.937	1.156	1.375	1.562	1.781	–
20 O.D.	20.0	–	0.250	0.375	0.500	0.375	0.593	0.812	0.500	1.031	1.281	1.500	1.750	1.968	–
22 O.D.	22.0	–	0.250	–	–	0.375	–	–	0.500	–	–	–	–	–	–
24 O.D.	24.0	–	0.250	0.375	0.562	0.375	0.687	0.968	0.500	1.218	1.531	1.812	2.062	2.343	–
26 O.D.	26.0	–	0.312	–	–	0.375	–	–	0.500	–	–	–	–	–	–
30 O.D.	30.0	–	0.312	0.500	0.625	0.375	–	–	0.500	–	–	–	–	–	–
34 O.D.	34.0	–	–	–	–	0.375	–	–	0.500	–	–	–	–	–	–
36 O.D.	36.0	–	–	–	–	0.375	–	–	0.500	–	–	–	–	–	–
42 O.D.	42.0	–	–	–	–	0.375	–	–	0.500	–	–	–	–	–	–

American Standard Graphical Symbols for Pipe Fittings and Valves

Item	Flanged	Screwed	Bell & Spigot	Welded	Soldered
Bushing					
Cap					
Cross					
Reducing					
Straight size					
Crossover					
Elbow					
45-degree					
90-degree					
Turned down					
Turned up					

Item	Flanged	Screwed	Bell & Spigot	Welded	Soldered
Base					
Double branch					
Long radius					
Reducing					
Side outlet (outlet down)					
Side outlet (outlet up)					
Street					
Joint					
Connecting pipe					
Expansion					

(continued)

American Standard Graphical Symbols for Pipe Fittings and Valves

Item	Flanged	Screwed	Bell & Spigot	Welded	Soldered
Lateral					
Orifice Flange					
Reducing Flange					
Plugs					
Bull plug					
Pipe plug					
Reducer					
Concentric					
Eccentric					
Sleeve					
Tee					
Straight Size					
Outlet up					

Item	Flanged	Screwed	Bell & Spigot	Welded	Soldered
Outlet down					
Double sweep					
Reducing					
Single sweep					
Side outlet (outlet down)					
Side outlet (outlet up)					
Union					
Angle Valve					
Check					
Gate (Elevation)					
Gate (Plan)					

(continued)

American Standard Graphical Symbols for Pipe Fittings and Valves

Item	Flanged	Screwed	Bell & Spigot	Welded	Soldered
Globe (Elevation)					
Globe (Plan)					
Hose Angle					
Automatic Valve					
By-pass					
Governor-operated					
Reducing					
Check Valve					
Angle check					
Straight way					
Cock					

Item	Flanged	Screwed	Bell & Spigot	Welded	Soldered
Diaphragm Valve					
Float Valve					
Gate Valve					
(Also used for general Stop valve symbol when amplified by specification)					
Angle gate					
Hose gate					
Motor-operated					
Globe Valve					
Angle globe					

(continued)

American Standard Graphical Symbols for Pipe Fittings and Valves

Item	Flanged	Screwed	Bell & Spigot	Welded	Soldered
Hose globe					
Motor-operated					
Hose Valve					
Angle					
Gate					
Globe					
Lockshield Valve					
Quick-opening valve					
Safety Valve					
Stop Valve					

American Standard Graphical Symbols for Piping

Category	*Symbol*
Air Conditioning	
Brine return	— — — BR — — —
Brine Supply	———— B ————
Circulating Chilled or hot-water flow	———— CH ————
Circulating chilled or hot-water return	— — — CHR — — —
Condenser water flow	———— C ————
Condenser water return	— — — CR — — —
Drain	———— D ————
Humidification line	—— - —— H —— - ——
Make-up water	—— - —— - —— - ——
Refrigerant discharge	———— RD ————
Refrigerant liquid	———— RL ————
Refrigerant suction	— — — RS — — —
Heating	
Air-relief line	—— ——— —— ——— ——
Boiler blow off	——— ——— ——— ———
Compressed air	———— A ————
Condensate or vacuum pump discharge	——o—— ——o—— ——o——
Feedwater pump discharge	——oo—— ——oo—— ——oo——
Fuel-oil flow	———— FOF ————
Fuel-oil return	— — — FOR — — —
Fuel-oil tank vent	— — — FOV — — —
High-pressure return	— // — // — // —
High-pressure steam	——//————//————//——
Hot-water heating return	— — — — — —
Hot-water heating supply	————————
Low-pressure return	— — — — — — —
Low-pressure steam	————————
Make-up water	—— - —— - —— - ——
Medium pressure return	— / — / — / —
Medium pressure steam	——/————/————/——
Plumbing	
Acid waste	ACID
Cold water	—— - —— - —— - ——
Compressed air	———— A ————
Drinking water flow	—— - —— - ——
Drinking water return	—— -- —— -- ——
Fire line	—— F —— F ——

(continued)

American Standard Graphical Symbols for Piping (Continued)

Category	Symbol
Gas	— G ———— G —
Hot water	
Hot water return	
Soil, waste, or leader (above grade)	
Soil, waste, or leader (below grade)	
Vacuum cleaning	— V ———— V —
Vent	
Pneumatic Tubes	
Tube runs	
Sprinklers	
Branch and head	
Drain	— S — — — S —
Main supplies	———— S ————

Weights and Measures

Commercial Weight
16 drams (dr) = 1 ounce (oz)
16 ounces = 1 pound (lb)
2000 pounds = 1 ton (t)

Dry Measure
2 pints (pt) = 1 quart (qt)
8 quarts = 1 peck (pk)
4 pecks = 1 bushel (bu)

Long Measure
12 inches (in) = 1 foot (ft)
3 feet = 1 yard (yd)
16 $\frac{1}{2}$ feet=1 rod (rd)
320 rd. (5280 ft.) = 1 mile (mi)

Time Measure
60 seconds (s) = 1 minute (min)
60 minutes = 1 hour (h)
24 hours = 1 day (d)
365 $\frac{1}{4}$ days=1 year (y)

Square Measure
144 square inches = 1 square foot
9 square feet = 1 square yard
$30\frac{1}{4}$ square yards = 1 square rod
$272\frac{1}{4}$ square feet = 1 square rod
40 square rods = 1 square rood
4 square roods = 1 square acre
43,560 square feet = 1 acre
640 acres = 1 square mile

Troy Weight
24 grains (gr) = 1 pennyweight (pwt)
20 pennyweights = 1 ounce (oz)
12 ounces = 1 pound (lb)
Circular Measure
60 seconds (") = 1 minute (′)
60 minutes (′) = 1 degree (°)
360 degrees (°) = 1 circle

Surveyors' Measure
7.92 inches (in) = 1 link (lk)
25 links ($16\frac{1}{2}$ ft) = 1 rod (rd)
4 rods (66 ft) = 1 chain (ch)
80 chains = 1 mile (mi)
Gunter's chain = 22 yards or 100 links
10 square chains = 1 acre

Liquid Measure
4 gills (gi) = 1 pint (pt)
2 pints = 1 quart (qt)
4 quarts = 1 gallon (gal)
$31\frac{1}{2}$ gallons = 1 barrel (bbl)

Cubic Measure
231 cubic inches = 1 gallon
2150.4 cubic inches = 1 bushel
1728 cubic inches = 1 cubic foot
27 cubic feet = 1 cubic yard
128 cubic feet = 1 cord (wood)
$24\frac{3}{4}$ cubic feet = 1 perch (stone)

Appendix B

Professional and Trade Associations

Table B-1 shows the addresses and Web sites for popular professional and trade associations.

Table B-1 Professional and Trade Associations

Organization	Address	Web Site
American Concrete Pipe Association	222 West Las Colinas Blvd.,Suite 641, Irving, TX 75039-5423	www.concretepipe.org
American Society of Mechanical Engineers	3 Park Avenue, New York, NY 10016	www.asme.org
American Society of Sanitary Engineers	901 Canterbury, Suite A Westlake, OH 44145	www.asse-plumbing.org
Environmental Protection Agency	1200 Pensylvania Avenue NW, Washington, DC 20460	www.epa.gov
International Association of Plumbing and Mechanical Officials	20001 East Walnut South, Walnut, CA 91789-2825	www.iapmo.org
The Institute of Plumbing	64 Station Lane, Hornchurch, Essex, RM12 6NB England	www.worldplumbing.org

(continued)

Table B-1 (continued)

Organization	Address	Web Site
Occupational Safety and Health Administration (U.S. Department of Labor)	200 Constitution Avenue NW, Washington, DC 20210	www.osha.gov
Plumbing and Drainage Institute	45 Bristol Drive, South Easton, MA 02375	www.pdionline.org
Plumbing, Heating, Cooling Information Bureau	222 Merchandise Mart Plaza, Chicago, IL 60654	www.phcib.org

Table B-2 shows popular Web sites where you can find important information.

Table B-2 Professional and Trade Associations

Topic	Web Site
Association acronyms	www.acronymfinder.com
Code-related Web sites	www.codesourcepc.com
Plumbing industry links	www.plumbinglinks.com

Appendix C

Exercise Answers

This appendix contains the answers to the exercises found at the end of chapters throughout this book.

Chapter 1
1. D
2. A
3. C
4. A
5. A
6. D
7. D
8. A
9. C
10. A

Chapter 2
1. B
2. B
3. B
4. D
5. B
6. B
7. D
8. B
9. A
10. D

Chapter 3
1. C
2. A
3. A
4. A
5. B

6. B
7. D
8. B
9. C
10. B

Chapter 4
1. N
2. O
3. M
4. I
5. L
6. K
7. F
8. H
9. G
10. E
11. D
12. B
13. C
14. A
15. J

Chapter 5
1. T
2. T
3. F
4. F
5. T

6. T
7. F
8. F
9. T
10. T

Chapter 6

1. C
2. A
3. D
4. A
5. A
6. A
7. C
8. A
9. B
10. A

Chapter 7

1. C
2. A
3. C
4. D
5. B
6. D
7. C
8. D
9. B
10. B

Chapter 8

1. C
2. A
3. B
4. C
5. A

6. D
7. C
8. D
9. A
10. B

Chapter 9

1. T
2. T
3. F
4. T
5. F
6. T
7. F
8. T
9. T
10. T

Chapter 10

1. A
2. C
3. B
4. D
5. A
6. A
7. C
8. B
9. D
10. A

Chapter 11

1. D
2. A
3. A
4. C
5. B

6. A
7. A
8. D
9. C
10. B

Chapter 12

1. sand
2. terrain
3. local
4. tamped
5. 850
6. third
7. $1/2$
8. prohibited
9. $1/4$
10. groundwater

Chapter 13

1. A
2. D
3. A
4. A
5. B
6. C
7. C

8. A
9. A
10. B

Chapter 14

1. A
2. B
3. B
4. C
5. D
6. B
7. C
8. B
9. C
10. C

Chapter 15

1. A
2. B
3. C
4. D
5. A
6. B
7. C
8. D
9. A
10. B

Glossary

Plumbing terms are defined here for the various city and state plumbing codes and examinations for licenses. The definitions that follow will not cover all of the terms, only the more pertinent ones.

Air break (drainage system). A drain piping arrangement from a fixture or device above the trap seal, but below the flood level rim.

Air gap. In a water supply system, the clear unobstructed vertical distance from any water supply pipe or faucet to a plumbing fixture or container and the flood level rim of the unit.

Alignment, pipe. Piping installed in a straight line, vertically, horizontally, or at a given angle.

Alloy steel. A steel with distinctive element properties other than carbon.

Approved. Work, materials, methods, and procedures that are acceptable by city, county, state, or federal approving authority for a given locality and a type of work or job.

Area drain. A receptacle or device designed and placed to collect surface water from a given open area.

Autopsy table. A table or unit used for postmortem medical examinations.

Backflow. The reverse flow of liquids in a pipe.

Backflow preventer. A device or means to prevent backflow into the potable water system.

Backpressure. Air pressure in plumbing pipes that is greater than the surrounding atmospheric pressure.

Back-siphonage. The flowing back by negative pressure of contaminated or polluted water from a plumbing fixture into a potable water system.

Backwater valve. A device installed in piping to prevent the back or reverse flow of storm or sewage into drainage systems or their branches.

Ball cock. A ball or device that opens or closes a faucet in accordance with a predetermined release procedure and floats on the surface of an enclosed container.

Bell (or hub). A pipe section that, for a short distance, has enlarged a portion of another pipe of identical diameter to form a joint.

Boiler blow-off. A controlled outlet on a boiler that permits discharge of sediment or the emptying of the unit.

Branch, plumbing. Any part of a piping system connected to a riser main or stack.

Branch vent. A vent that connects one or more vents with a stack vent.

Building. Any structure designed and built for support, shelter, recreation, and enclosure for persons, animals, or property when considered for plumbing installation procedures.

Building drain. At the lowest horizontal point of the building, a drain that receives soil or waste from the interior of the building and conveys it to an approved point of discharge.

Building storm drain. That part of the piping of the building drainage system that takes surface water, groundwater, subsurface water, cooling water, or similar discharge to a public or approved discharge point.

Building sub-drain. That portion of a building drainage system that cannot drain by gravity into the building drainage system.

Burr. Protruding metal or roughness on the walls of a pipe resulting from pipe cuts.

Carbon steel. Steel that contains a high percentage of carbon as distinguished from the other elements.

Cesspool. A line or covered excavation in the ground that receives wastes from the drainage system and retains organic matter and solids. The liquids seep into the ground.

Circuit vent. A vent designed to serve two or more traps extending from in front of the last fixture connection of a horizontal branch to the vent stack.

Cistern. A covered (usually small) tank used chiefly for storing rainwater for domestic use other than for drinking purposes.

Cleanout. A metallic plug or cover that can be removed for the purpose of cleaning or examining the interior of the pipe. It is joined by means of a screw thread to an opening in a pipe.

Code. Regulations adopted by an administrative agency that has jurisdiction for a particular locality.

Combination fixture. A fixture that combines a sink and laundry tray or two or three compartment laundry trays into one unit.

Combined building sewer. A sewer that receives both sewage and storm water.

Common or continuous waste. A waste with several compartments (such as a double laundry tray) connected to a single trap.

Common vent (dual vent). A vent that serves two fixtures by a connection at the junction of the two fixture drains.

Compressive stress. A stress that resists a force attempting to crush a body.

Conductor or leader. Part of a roofing and/or area gutter system that takes water from a roof or above-surface area to a storm drain or other disposal area or system.

Dead end. The extended portion of a pipe that is closed at one end to which no connections are made. The dead end is without circulation of free air.

Developed length. The total length of a pipeline measured along the centerline and the fittings of the pipe.

Diameter. The nominal commercial designation, normally the inside diameter of the pipe (unless otherwise specifically stated in a particular plumbing code).

Drainage system. All the drainage piping within a building that conveys sewage, rainwater, and other liquid wastes to a legal point of disposal exclusive of public and private exterior pipelines or systems.

Durham system. A system of piping in which all piping is of a thread type with recessed drainage fittings.

Ejector. An electrically or mechanically operated device used to elevate sewage and liquid wastes from a lower level to a point of discharge into a sewer or other disposal system.

Elastic limit. The greatest stress a material can withstand without permanent deformation after stress release.

Ferrule. A metallic sleeve or fitting used to connect dissimilar plumbing materials.

Fixture branch. A water supply source serving more than one fixture.

Fixture supply. A water supply source connecting the fixture and supply pipe.

Fixture unit. A flow of waste, used for sizing plumbing jobs in a plumbing system, equal to 1 cubic foot per minute.

Float valve. A valve used to control the water level in a tank or other container. It is operated by a float and is considered a positive operating valve.

Flood level. The level at which water begins to overflow the top or rim of a plumbing fixture.

Flush valve. A valve used for flushing water closets and similar fixtures.

Grade (gradient). The fall or slope of a line of pipe in reference to a horizontal plane. In drainage systems, it is usually expressed as the fall in a fraction of an inch per foot length of pipe.

Grease interceptor. A container designed to intercept and hold grease or fatty substances (portions of food waste) from a line to which it is connected.

Horizontal pipe. An installed pipe or fitting that forms an angle of more than 45° with the vertical.

Individual vent. A vent installed in a pipe to vent a fixture trap connected to the vent system above the fixture it serves.

Invert. The floor or lowest part of the internal cross section of a conduit or pipe.

Main. The principal pipe of a system of continuous piping to which branch connections can be made.

Main vent. The principal vent pipe of a system to which vent connections can be made.

Manhole. An opening in a plumbing system or sewer large enough to permit a person to gain access to the system for purposes of inspection and/or cleaning.

Plumbing system. All water supply, drainage, and venting systems and all fixtures and their traps, complete with their connections.

Potable water. Water that meets the standards of a government agency and is used for culinary, domestic, and drinking purposes.

Receptor. A receptacle designed to receive discharges from indirect waste piping.

Relief vent. A vent that provides circulation of air between the vent and drainage system.

Roof drain. A drain that receives water collected on a roof surface and discharges it into a collection unit or downspout.

Roughing-in. The installation of all parts of a plumbing system or a particular unit that can be completed before the installation of fixtures. This includes soil, vent, waste, and water supply piping and supports.

Sanitary sewer. A sewer that carries only sewage wastes and excludes storm water, surface water, and groundwater.

Septic tank. A watertight receptacle that receives sewage.

Service pipe. The water supply line from the source of supply to the building served.

Sewage. Liquid waste that contains vegetable or animal matter in suspension or solution including liquids containing chemicals in solution. These liquids are the wastes being carried away from public and industrial buildings and from residences.

Shearing stress. A stress that resists a force that would make one layer of a body slide across another layer.

Siphonage. A suction resulting from the flow of liquids in pipes.

Slip joint. A connection in which one pipe slides into another to make a tight joint with a threaded retainer or an approved gasket.

Soil pipe. A pipe that carries the discharge of toilets or similar fixtures (with or without discharges of other fixtures) to a sewer or approved drain.

Stack vent. The vertical extension of soil, waste, or vent system above the highest horizontal drain connected to the stack.

Strain. The change of size or shape of a body produced by the action of a stress.

Stress. The reactions of a body to external forces (in this case, metals).

Sump. A tank or pit below the normal grade of gravity receiving liquid wastes or sewage from which the wastes or sewage must be mechanically pumped to a higher level receiving point.

Supports, hangers, anchors. Devices used for supporting or securing fixtures, piping, and fittings to ceilings, floors, or walls.

Temperature-pressure relief valve. A valve installed on top of the hot-water heater tank to relieve the buildup of dangerous temperatures or pressure inside the tank should its heating system fail to turn off automatically.

Tensile strength. The overall tensile stress that a material will develop. The tensile strength is considered to be the load in pounds per square inch at which test materials rupture.

Tensile stress. A stress that resists a force that tends to pull a body apart.

Trap. A designed fitting normally shaped with a U-type part that continuously retains a liquid seal and prevents the back passage of air without affecting the flow of liquids in the system.

Tuberculation. The process whereby steel and iron pipes rust.

Turbulence. The deviations from parallel flow in a pipe because of rough inner walls, obstructions, or directional changes.

Vacuum breaker. A device designed to protect a water supply system against back-siphonage by providing an opening through which air may be drawn to relieve negative vacuum pressure.

Vent pipe. A pipe or system of pipes provided to ventilate a plumbing system.

Volume of a pipe. The measurement of the space within pipe walls. To find the volume of a pipe, multiply the length or height of the pipe by the product of pi and the square of the inside radius.

$$0.7854\left[\frac{\pi}{4}\right]$$

This is written as

$$V = \frac{\pi}{4} \times D^2$$

Wet vent. Any drainpipe that in addition to receiving discharge waste also serves as a vent.

Wiped joint. The adhesion of metal with solder at a finish thickness of at least $1/4$ inch at the point where the pipes are joined. The joint shall be smoothly finished with a wiping cloth.

Workmanship. Work of an acceptable nature and character that will fully secure the expected results of the locality codes for the safety, welfare, and health of all persons.

Yoke vent. A pipe from a soil or waste stack with an upward connection to a main vent pipe for purposes of preventing pressure variances in the stack.

Index